Electricity Markets and Power System Economics

Electricity Markets and Power System Economics

Deqiang Gan

Donghan Feng

Jun Xie

CRC Press
Taylor & Francis Group
Boca Raton London New York

CRC Press is an imprint of the
Taylor & Francis Group, an **informa** business

CRC Press
Taylor & Francis Group
6000 Broken Sound Parkway NW, Suite 300
Boca Raton, FL 33487-2742

Version Date: 20130916

International Standard Book Number-13: 978-1-4665-0169-0 (Hardback)

Library of Congress Cataloging-in-Publication Data

Gan, Deqiang, 1966-
    Electricity markets and power system economics / Deqiang Gan, Donghan Feng, Jun Xie.
        pages cm.
    Includes bibliographical references and index.
    ISBN 978-1-4665-0169-0 (hardback)
    1. Electric utilities. 2. Power resources--Finance--Mathematical models. I. Feng, Donghan, 1966- II. Title.

HD9685.A2G363 2014
333.793'2--dc23   2013028522

Visit the Taylor & Francis Web site at
http://www.taylorandfrancis.com

and the CRC Press Web site at
http://www.crcpress.com

Contents

Preface

The notion of electricity marketing is not new at all. After the first power plant in history was commissioned for commercial operation by Thomas Edison on Pearl Street in New York in 1882, electricity was sold as a consumer product at market prices. After a period of rapid development, electricity became such a fundamental product that regulation was believed to be necessary. Since then, the power industry had been considered a natural monopoly and undergone periods of tight regulation. Deregulation started in the early 1980s and as a result, most developed countries run their power industries using a market approach.

The practices involved in marketing electricity markets change quickly. Market rules are published every year, then updated and published again. Additionally, relevant publications and research reports are reviewed and archived. The need to describe the basic building blocks of electricity market theory systematically is obvious and this book was written to meet the need. After reading this book, an electric power professional should understand the ramifications of mainstream market rules and be able to read scholarly articles.

About half the material covered in the book came from our research results. However, the book is more of a textbook and clarity is the top priority when writing a text. The intended audiences are senior undergraduate students and first-year graduate students. Chapters 1 through 3 provide a basis for understanding the rest of the book. The remaining chapters can be read almost independently. Figure 0.1 depicts the dependency of the chapters in this book.

We thank our co-workers for doing a great job in performing the research described in this book and for their generosity in allowing us to report their research results. The co-workers are named in the papers we cited in each chapter.

Deqiang Gan
Zhejiang University

Donghan Feng
Shanghai Jiao Tong University

Jun Xie
Nanjing University of Posts and Telecommunications

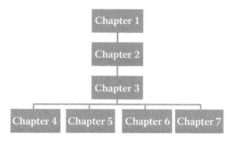

Figure 0.1   Dependency of chapters in the book.

Chapter 1

Introduction

This chapter briefly introduces the basic concepts of microeconomics. By explaining terms such as supply, demand, and market equilibrium that appear frequently in this book, we aim to prepare readers who have no economics background to understand all the economic aspects of electricity marketing. Readers are referred to Pindyck and Rubinfeld (1997) at the undergraduate level or Luenberger (1995) at the graduate level for more details. Later in this chapter, we summarize the history of the development of electricity markets.

1.1 Demand and Supply

In a modern society, people everywhere participate in markets by creating demands and buying commercial products. Every buyer exerts preferences by buying certain products. As a result, the total demand for a product is determined by the preferences of consumers. Another important (and obvious) factor that affects total product demand is price. When studying demand functions, we often view demand for a product strictly as a function of price.

As an example, we denote electrical power demand as $P(\rho)$; ρ represents the price of power. It makes perfect sense to assume that the power demand function $P(\rho)$ decreases monotonically, as Figure 1.1 shows.

In subsequent text we often mention the concept of the inverse demand function $\rho(P)$. Figure 1.2 depicts a typical inverse demand function. On the other side of a market, the behaviors of suppliers can be described using a supply function that shows the relationship of price and quantity supplied (Figure 1.3).

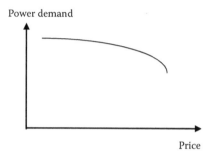

Figure 1.1   Relationship between demand and price.

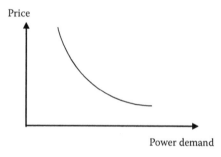

Figure 1.2   Typical inverse demand function.

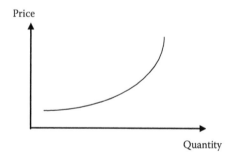

Figure 1.3   Relationship of price and quantity supplied.

1.2 Market Equilibrium

The concept of market equilibrium is central to microeconomics. Figure 1.4 shows an equilibrium point at which the demand and supply curves meet (Figure 1.4). Point P denotes the market clearing price. If the market price is $P1 > P$, a surplus of supply exists. This surplus forces market suppliers to lower prices, pulling market prices back to the equilibrium point. When the market price equals $P2 < P$,

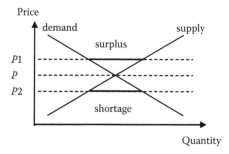

Figure 1.4 Supply and demand.

a supply shortage exists and causes suppliers to increase prices, pushing market prices back to the equilibrium point.

We explained that market surpluses and shortages stabilize market prices and act as "invisible hands" behind a free market. Furthermore, at the market equilibrium point, both demand and supply enjoy certain levels of surplus (Figure 1.5) for this reason. Since the market price is equal to P, certain consumers are willing to pay more than P (say, $P1 > P$). The area of the triangle above line $P = 0$ is a measure of demand surplus. Similarly, a supply surplus exists and the area of the triangle beneath the line $P = 0$ is a measure of supply surplus.

A government may regulate or intervene in a market by setting a price ceiling on a product. Let the price ceiling designated $P2$ (Figure 1.6) push the supply side to lower the production level to $Q2$. We can see from the figure that the demand surplus varies by the area A − B while the supply surplus varies by −A − C. The end result of this regulation policy is that social welfare is decreased by the area B + C. A standard term for this inefficiency caused by government regulation is *deadweight loss*.

What if the government sets a minimum price instead of a price ceiling? Let the minimum price level be $P1$. As a result of government intervention, the demand drops to $Q1$ (Figure 1.7). It can be seen that the demand surplus varies by −A − B,

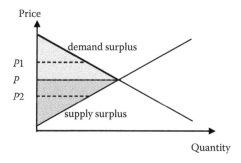

Figure 1.5 Demand surplus and supply surplus.

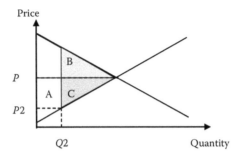

Figure 1.6   Setting a price ceiling on a product.

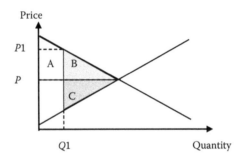

Figure 1.7   Setting a minimum price on a product.

while the supply surplus varies by A − C. The end result is also unsatisfactory: social welfare is lowered by B + C. This equilibrium analysis shows that government intervention (the "visible hand") usually introduces a social welfare loss, but this assertion may not apply to all markets.

1.3 Price Elasticity and Competitive Markets

The familiar concept of demand elasticity has a precise mathematical definition:

$$\varepsilon(\rho) = \frac{\rho}{P} \frac{dP}{d\rho} \tag{1.1}$$

We are not very interested in the specific quantity of demand elasticity in this book; rather we are interested only in two extremes. At one extreme, the demand for a certain product can be very elastic to the price of the product. For such products, the inverse demand functions are horizontal, as shown in Figure 1.8.

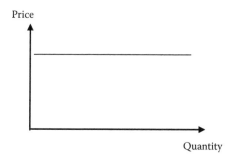

Figure 1.8   Demand is very elastic to price.

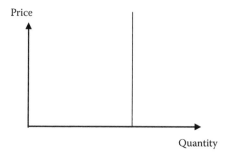

Figure 1.9   Demand is very inelastic to price.

At the other extreme, demand can be very inelastic to price. Examples of products in this category include food, clothing, and, of course, electricity. The inverse demand functions for such products are vertical (Figure 1.9). In this book we are mainly concerned with inverse demand functions.

A commercial product incurs costs for raw materials. A simple example is electricity generation that consumes gas, coal, or another form of primary energy. In general, the cost of a product is a function of the amount of raw materials required and the production quantity. Here we assume that the cost of a product is a function of production quantity only. In particular, the cost function of generating electricity is denoted as $c(P)$. The marginal cost of production is defined as dc/dP. Figure 1.10 depicts the production function.

Now we are in a position to introduce an important result of production theory: the *fundamental law of production*. Understandably, the objective of a firm is to maximize its profit. Suppose that the inverse demand function of a firm's product is $\rho(P)$. The profit maximization problem is formulated as

$$\underset{P}{Max} \quad \pi(P) = v(P) - c(P) = \rho(P)P - c(P) \tag{1.2}$$

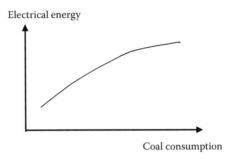

Figure 1.10 Production function.

where $v(P)$ is the return on the product. Differentiating with respect to P, we obtain:

$$\frac{dv}{dP} = \frac{dc}{dP} \qquad (1.3)$$

which states that profit maximization is achieved when the marginal cost of a product is equal to its marginal profit. Notice that $v(P) = \rho(P)P$, so

$$\frac{dv}{dP} = \rho(P) + \frac{d\rho}{dP} P.$$

The fundamental law of production can be reformulated as

$$\rho(P) + \frac{d\rho}{dP} P = \frac{dc}{dP} \qquad (1.4)$$

Recall the definition of price elasticity:

$$\varepsilon = \frac{\rho}{P}\frac{dP}{d\rho}.$$

Thus we have

$$\frac{d\rho}{dP} = \frac{\rho}{P\varepsilon}.$$

Substituting it into Equation 1.4, we find:

$$\rho = \frac{\dfrac{dc}{dP}}{\left(1 + \dfrac{1}{\varepsilon}\right)} \tag{1.5}$$

In general, $dP/d\rho < 0$, which implies $\varepsilon < 0$. Equation 1.5 points the way to a pricing strategy. If the price elasticity of a product is low, the corresponding price should be high; if the elasticity is high, the price should be low. This is the outcome we usually see in daily life. Now let us consider an idealized situation in which product elasticity is extremely high. Under such a situation, the inverse demand function $\rho(P)$ is a horizontal line, so $1/\varepsilon = 0$. As a result Equation 1.5 simplifies to

$$\rho = \frac{dc}{dP} \tag{1.6}$$

This indicates that the market of a company facing a horizontal inverse demand function is perfectly competitive because the product price equals the product marginal cost. Mainstream economists believe that such a market yields maximum social welfare. Mas-Colell et al. (1995) explained the fundamental theorem of welfare economics under mild conditions: "If the price and quantity of a market constitute a competitive equilibrium, then this allocation is Pareto optimal." We will later use a game-theoretic method to show that as the number of suppliers in a market approaches infinity, the market becomes perfectly competitive.

1.4 Economy of Scale and Natural Monopoly

A production function describes the relationship between the inputs and outputs of a firm or industry. Using a coal power plant as an example, let z be the coal consumption level, E denote the electrical energy produced, and $E(z)$ indicate the production function of the firm. Production functions generally increase monotonically.

We make two remarks here. First, it is often assumed that a firm or an industry always chooses the most economic means available to organize its production. For instance, a power company always uses an optimization solution to dispatch its units. The second remark about production function is relevant to our discussion because it concerns the scale and economics of production. If for any $t > 0$ and z, we have $E(tz) > tE(z)$, we say that the production enjoys economy of scale; the opposite function demonstrates diseconomy of scale (Figure 1.11).

We can also determine the scale property of a production function by looking at the marginal cost function. If a production function decreases monotonically,

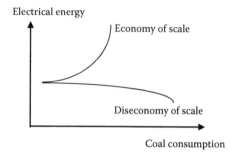

Figure 1.11 Economy of scale and diseconomy of scale.

the production exhibits economy of scale; if it increases monotonically, the reverse applies.

We conclude this part of the text with a fundamental concept of modern microeconomics: *an industry with economy of scale cannot achieve perfect competition using a market approach*. Such an industry is called a natural monopoly and invites government regulation.

1.5 Brief History of Electricity Markets

For a long period, economists believed that the electrical energy industry enjoyed an economy of scale and was thus considered a natural monopoly. In fact, the power industry faced and continues to face a nearly vertical demand. Government regulation had been applied and the industry operated in a centralized fashion using the common rate-of-return scheme.

Advances in generation technology changed the picture. In the early 1980s, the degree of economy of scale decreased to a level that competition on the generation side of the industry made sense. However, the transmission sector remains a natural monopoly worldwide. The following list summarizes important events in the development of electricity markets:

- 1978: The Public Utility Regulatory Policies Act (PURPA) was approved in the United States. It allowed non-utility generators to receive reasonable prices for energy they produced.
- 1982: Chile started running the first modern spot market. Fred Schweppe published a famous paper (Schweppe 1982) setting the foundation of nodal pricing now widely adopted in electricity markets.
- 1990: The United Kingdom established a pool-based electricity market that served many nations for many years as a market model.
- 1994: The Nordic electricity market started functioning and constituted the first multi-national spot market.

- 1996: Spot electricity markets appeared in Australia and New Zealand; California established a spot market based on zonal pricing—a coarse version of nodal pricing.
- 1998: A spot market was created in Pennsylvania, New Jersey, and Maryland and also served as a worldwide market model. Later, the New England region and New York State followed a similar market form.
- 2001: The United Kingdom completed a second market reform featuring bilateral contract-based free trading.

Looking into the future, the industry still faces many technical challenges. To mention just one, short-term electrical demand is still highly inelastic. An immediate consequence is the lack of an effective pricing system that introduces a fundamental limitation to the overall efficiency of electricity markets. Readers should read the three review articles (Green 2000; Joskow 1997; Sweet et al. 2002) listed in the reference section.

References

M.C. Caramaris, R.E. Bohn, F.C. Schweppe. 1982. *Optimal Spot Pricing: Practice and Theory.* IEEE Trans. On Power Apparatus and Systems, PAS-101, 2234–3245.

R.J. Green. 2000. Competition in generation: the economic foundations. *Proceedings of IEEE*, 88, 128–139.

P.L. Joskow. 1997. Restructuring, competition and regulatory reform in the U.S. electricity sector. *Journal of Economic Perspectives*, 11, 119–138.

D. Luenberger. 1995. *Microeconomic Theory.* New York: McGraw Hill.

A. Mas-Colell, M. Whinston, and J. Green. 1995. *Microeconomic Theory.* Oxford: Oxford University Press.

R.S. Pindyck and D.L. Rubinfeld. 1997. *Microeconomics.* New York: Prentice Hall.

W. Sweet, E.A. Bretz, and A. Kahn. 2002. How to make deregulation work. *IEEE Spectrum*, 39.

Chapter 2

Fundamentals of Power System Operation

The physical constraints of modern power systems set electricity markets apart from other traditional markets. This chapter provides an overview of the basic operational concepts of power systems, such as load flow, spinning reserve, transmission capability, etc. The Karush–Kuhn–Tucker conditions are also introduced as the mathematical basis for analyzing the optimization problems in this book. Most of the materials covered in this chapter are standard and they can be found in the references at the end of this chapter.

Generally speaking, a real-time electricity market is indispensable for most electricity markets operating around the world. Therefore, to learn electricity market theory, we first should study the elementary theories of power system operation.

2.1 Economic Dispatch

The most important elements in a power system are generators that make use of the primary energy form coal, oil, gas, nuclear, or hydroelectric power. The costs for operating generators can be classified as fixed and variable. The operation of a power system should consider both kinds of costs. Generally, after a generator is commissioned, its fixed cost can be allocated to variable cost and engineers can devise generator cost curves. Let P_G denote the generator's output. Its cost curve can be expressed by using function $c(P_G)$.

For a forecasted load, the generators should be dispatched in a way to minimize operation cost. Cost minimization is the basis for economic dispatch. The mathematical model for this problem can be expressed as an optimization problem:

$$\underset{\mathbf{P}_G}{Min} \quad \sum_{i=1}^{NG} c_i(P_{Gi}) \tag{2.1}$$

$$S.T. \quad \mathbf{e}^T \mathbf{P}_G = P_D \tag{2.2}$$

$$\underline{\mathbf{P}}_G \leq \mathbf{P}_G \leq \overline{\mathbf{P}}_G \tag{2.3}$$

where $\mathbf{e}^T = [1 \ 1, \ldots, 1]$ $\underline{\mathbf{P}}_G$ and $\overline{\mathbf{P}}_G$ are the maximum and minimum power outputs of generators, respectively; P_D is the total system load demand; and NG is the number of generators. Now let us explain how to solve the above optimization problem. In this section, we consider a condition that has no inequality constraints. The problem is then reduced to

$$\underset{\mathbf{P}_G}{Min} \quad \sum_{i=1}^{NG} c_i(P_{Gi}) \tag{2.4}$$

$$S.T. \quad \mathbf{e}^T \mathbf{P}_G = P_D \tag{2.5}$$

The above problem can be solved by using the classical Lagrange multiplier method. Let λ denote the Lagrange multipliers that correspond to equality constraints; then the above equality-constrained optimization problem can be changed to

$$\underset{\mathbf{P}_G, \lambda}{Min} \quad \Gamma = \sum_{i=1}^{NG} c_i(P_{Gi}) + \lambda(\mathbf{e}^T \mathbf{P}_G - P_D) \tag{2.6}$$

Differentiating with respect to $\mathbf{P}_G$ and λ, we obtain the optimization condition:

$$\begin{cases} \dfrac{\partial \Gamma}{\partial P_{Gi}} = \dfrac{\partial c_i}{\partial P_{Gi}} + \lambda = 0, \quad i = 1, 2, \ldots, NG \\[4mm] \dfrac{\partial \Gamma}{\partial \lambda} = \mathbf{e}^T \mathbf{P}_G - P_D = 0 \end{cases} \tag{2.7}$$

The above formula constitutes the so-called coordination equations. According to the coordination equations, the optimal solution for economic dispatch is to dispatch the output of generators so that the incremental costs of all the units are equal. This is based on the fact that the optimal solution satisfies the follow equation:

$$\frac{\partial c_1}{\partial P_{G1}} = \frac{\partial c_2}{\partial P_{G2}} = \ldots = \frac{\partial c_{NG}}{\partial P_{G,NG}} = -\lambda \qquad (2.8)$$

The above result is the famous law of equal incremental rates.

2.2 Load Flow Calculation

Transmission lines will burn out if the electric current passing through them is too large. Therefore, the transmission capability of a system is not infinite. When dispatching generators, operators must ensure that the transmission line power flow will not overload. For example, in the power system shown in Figure 2.1, the power output of generator G1 should be restricted below 150 MW, although its bid price is lower than generator G2. The bid price of generator G2 is high but its power output must be bounded above 50 MW.

It is more complicated to calculate load flow for a practical power system than for a two-generator system. The basic method is built on the electromagnetic field and electric circuit theories. Simply stated, a transmission line model can be built based on electromagnetic field theory. Generators and loads are represented using constant nodal injection powers. The combined non-linear equations of transmission lines and nodes are based on electric circuit theory. The state of a power system at a certain time can be described by an electric circuit model. The state variables include the complex voltage and current at each bus, as shown in Figure 2.2. The meanings of the variables in Figure 2.2 are as follows:

$P_{ij} + jQ_{ij}$	power injection at node i
i'	current on branch to earth at node i
$R_{ij} + jX_{ij}$	transmission line impedance
I_{ij}	current injection at node i
$R_{io} + jX_{io}$	impedance on branch to earth at node i
$R_{jo} + jX_{io}$	impedance on branch to earth at node j
i''	current in transmission line

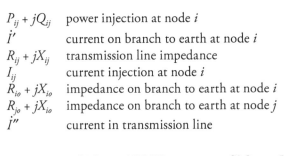

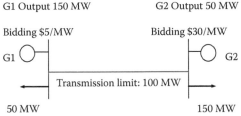

G1 Output 150 MW   G2 Output 50 MW

Bidding \$5/MW   Bidding \$30/MW

G1   G2

Transmission limit: 100 MW

50 MW   150 MW

Figure 2.1 Two-generator power system.

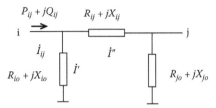

Figure 2.2 Electric circuit model for power system components.

Let $\dot{\mathbf{I}}$ and $\dot{\mathbf{V}}$ denote node circuit and node voltage complex vector, respectively. $\mathbf{Y}$ is the node admittance matrix. The power system state variables (voltage and circuit) satisfy Kirchhoff laws:

$$\dot{\mathbf{I}} = \mathbf{Y}\dot{\mathbf{V}} \tag{2.9}$$

Let N be the total number of nodes. The above formula can be formulated as

$$
\begin{bmatrix} \dot{I}_1 \\ \dot{I}_2 \\ \cdots \\ \dot{I}_N \end{bmatrix}
=
\begin{bmatrix}
Y_{11} & Y_{12} & \cdots & Y_{1N} \\
Y_{21} & Y_{22} & \cdots & Y_{2N} \\
\cdots & \cdots & \cdots & \cdots \\
Y_{N1} & Y_{N2} & \cdots & Y_{NN}
\end{bmatrix}
\begin{bmatrix} \dot{V}_1 \\ \dot{V}_2 \\ \cdots \\ \dot{V}_N \end{bmatrix}
\tag{2.10}
$$

Let

$$y_{ij} = \frac{1}{R_{ij} + jX_{ij}} \tag{2.11}$$

The off-diagonal element Y_{ij} in node admittance *matrix $i \neq j$* is defined as

$$Y_{ij} = -y_{ij} \tag{2.12}$$

The definition of diagonal elements of the node admittance matrix is

$$Y_{ii} = \sum_{\substack{j \\ j \neq i}} y_{ij} \tag{2.13}$$

Notice that

$$S = P_i + jQ_i = \dot{V}_i \hat{I}_i = \dot{V}_i \sum_{j=1}^{N} \hat{Y}_{ij} \hat{V}_j \tag{2.14}$$

Therefore

$$P_i + jQ_i - \dot{V}_i \sum_{j=1}^{N} \hat{Y}_{ij} \hat{V}_j = 0 \tag{2.15}$$

The node voltage is expressed in polar coordinates $\dot{V} = V \angle \theta$. The voltage vector in the above equation is unknown. Meanwhile, the node admittance matrix is a constant and relates to transmission line parameters and system topology. Generally, we know the node real power injection, node real power load, node reactive power injection, and node reactive power load. We expand the above equation and rewrite it as

$$\begin{cases} P_{Gi} - P_{Di} - V_i \sum_{j=1}^{N} \left(G_{ij} V_j \cos \theta_{ij} + B_{ij} V_j \sin \theta_{ij} \right) = 0 \\ \\ Q_{Gi} - Q_{Di} - V_i \sum_{j=1}^{N} \left(G_{ij} V_j \sin \theta_{ij} - B_{ij} V_j \cos \theta_{ij} \right) = 0 \end{cases} \tag{2.16}$$

where P_{Gi} and P_{Di} are the generator's real power output and load at node i, respectively, and are constant in load flow equations. Q_{Gi} and Q_{Di} are the generator's reactive power output and load at node i, respectively, and they are also constant; G and B are the real part and imaginary part of the node admittance matrix, respectively; and V_i and θ_i are the complex voltage module and phase angle at node i, respectively, $\theta_{ij} = \theta_i - \theta_j$.

In this chapter, we are concerned mainly with real power problems and reactive powers are omitted. In what follows, we derive the equations for real power flow only. It is often true that the voltage magnitude is known or can be estimated roughly in advance (in practice, it changes little), and the voltage phase angle differences between the two ends of transmission lines are small. Therefore, $\sin \theta_{ij} \approx \theta_{ij}$, $\cos \theta_{ij} \approx 1$. Meanwhile, let $G_{ij} \approx 0$. Then we can simplify the above load flow equations and obtain the following linear equations:

$$P_{Gi} - P_{Di} - V_i \sum_{j=1}^{N} V_j B_{ij} \theta_{ij} = 0 \tag{2.17}$$

The above real power load flow equation is implemented in many power dispatch centers. If the voltage equals 1.0, we obtain an even simpler form of load flow equation:

$$P_{Gi} - P_{Di} - \sum_{j=1}^{N} B_{ij}\theta_{ij} = 0 \qquad (2.18)$$

In the above formula,

$$\sum_{j=1}^{N} B_{ij}\theta_{ij}$$

can be reformulated as

$$\theta_i \sum_{j=1}^{N} B_{ij} - \sum_{j=1}^{N} B_{ij}\theta_j$$

If the branch reactance is ignored when forming the admittance matrix (based on the definition of admittance matrix),

$$\sum_{j=1}^{N} B_{ij} = 0$$

Therefore, the above equation can be reformulated as

$$P_{Gi} - P_{Di} + \sum_{j=1}^{N} B_{ij}\theta_j = 0 \qquad (2.19)$$

or

$$P_{Gi} - P_{Di} = \sum_{j=1}^{N} (-B_{ij})\theta_j \qquad (2.20)$$

The above equation can be reformulated in the form of a matrix:

$$\mathbf{P}_G - \mathbf{P}_D = \mathbf{B}\theta \qquad (2.21)$$

Therefore, we obtain a concise form of a DC load flow equation. Now we can find an approximate expression for branch load flow. First, let us review the branch model shown in Figure 2.2, in which $G_{ij} + jB_{ij}$, $G_{io} + jB_{io}$, and $G_{jo} + jB_{jo}$ denote branch admittance $i \neq j$. Let $\hat{I}$ denote the conjugate of current $P_{ij} + jQ_{ij}$ and indicate the complex power injection at node i. Then, according to general electric circuit theory:

$$
\begin{aligned}
P_{ij} + jQ_{ij} &= \dot{V}_i \hat{I}_{ij} = \dot{V}_i(\hat{I}' + \hat{I}'') \\
&= \dot{V}_i[\hat{V}_i(G_{i0} - jB_{i0}) + (\hat{V}_i - \hat{V}_j)(G_{ij} - jB_{ij})] \\
&= \dot{V}_i(\hat{V}_i G_{j0} - j\hat{V}_i B_{i0} + \hat{V}_i G_{ij} - j\hat{V}_i B_{ij} - \hat{V}_j G_{ij} + j\hat{V}_j B_{ij})
\end{aligned}
\tag{2.22}
$$

Notice that $\dot{V} = V^{j\theta}$. Let $\theta_i - \theta_j = \theta_{ij}$; therefore

$$
\begin{aligned}
P_{ij} + jQ_{ij} &= \left[(G_{i0} + G_{ij})V_i^2 - (B_{ij} \sin\theta_{ij} + G_{ij} \cos\theta_{ij})V_i V_j\right] \\
&\quad + j\left[-(B_{i0} + B_{ij})V_i^2 + (B_{ij} \cos\theta_{ij} - G_{ij} \sin\theta_{ij})V_i V_j\right]
\end{aligned}
\tag{2.23}
$$

Hence,

$$
P_{ij} = (G_{io} + G_{ij})V_i^2 - (B_{ij} \sin\theta_{ij} + G_{ij} \cos\theta_{ij})V_i V_j
\tag{2.24}
$$

As an estimate, let $V_i = V_j = 1$, $\sin\theta_{ij} \approx \theta_{ij}$, $G_{io} \approx 0$, and $G_{ij} \approx 0$; then

$$
P_{ij} = -B_{ij}\theta_{ij}
\tag{2.25}
$$

Generally, we let F_{ij} denote transmission line real power flow. Since

$$
B_{ij} = \frac{-X_{ij}}{R_{ij}^2 + X_{ij}^2}
$$

we set the branch resistance at zero; then

$$
B_{ij} = -\frac{1}{X_{ij}}
$$

where X_{ij} is the reactance of line $i - j$. The load flow F_{ij} in line $i - j$ can thus be calculated:

$$
F_{ij} = (\theta_i - \theta_j)/X_{ij}
\tag{2.26}
$$

Let **C** indicate the node coincidence matrix. The structure of **C** appears as follows:

$$
\mathbf{C} = \begin{bmatrix} \cdots & & & & \\ \cdots & 1 & \cdots & -1 & \cdots \\ \cdots & & & & \end{bmatrix} \begin{array}{l} \\ \text{line } i - j \\ \\ \end{array}
\qquad (2.27)
$$

i row $\qquad$ j row

When forming the above matrix, we suppose the load flow is from node i to node j. Let matrix **X** have the same structure as matrix **C**. The definition of the elements in matrix **X** is

$$
\mathbf{X} = \begin{bmatrix} \cdots & & & & \\ \cdots & \dfrac{1}{X_{ij}} & \cdots & -\dfrac{1}{X_{ij}} & \cdots \\ \cdots & & & & \end{bmatrix} \begin{array}{l} \\ \text{line } i - j \\ \\ \end{array}
\qquad (2.28)
$$

i row $\qquad$ j row

Then the line load flow equation can be rewritten in the form of a matrix as follows:

$$
\mathbf{F} = \mathbf{X\theta} \qquad (2.29)
$$

For computational purposes, sparse matrices should be employed for the above power flow model. Later in this chapter, we will derive a dense formulation of the model to study economic dispatch problems. Note that the inverse of matrix **B** is non-existent when solving DC load flow equations. We generally select a reference node and set the voltage angle of the node to 0. For ease of illustration, we assume a system contains N nodes and the reference is node N. We use the variables that have a prime sign (′) to denote the vector and matrix after deleting the reference node-related terms. Note that when nodal injection power vector $\mathbf{P}_G'$ and load vector $\mathbf{P}_D'$ are known, the net power injection of the reference node equals $\mathbf{e}^T (\mathbf{P}_G' - \mathbf{P}_D')$. The direct current load flow equation can be reformulated as

$$
\begin{cases} \mathbf{P}_G' - \mathbf{P}_D' = \mathbf{B}'\mathbf{\theta}' \\[2mm] P_{GN} - P_{DN} = -\mathbf{e}^T (\mathbf{P}_G' - \mathbf{P}_D') \\[2mm] \theta_N = 0 \end{cases}
\qquad (2.30)
$$

The voltage angles of buses can be calculated as

$$\boldsymbol{\theta}' = \mathbf{B}'^{-1}(\mathbf{P}'_G - \mathbf{P}'_D) \tag{2.31}$$

Notice that

$$\mathbf{F} = \mathbf{X}\boldsymbol{\theta} = \mathbf{X}\begin{bmatrix} \mathbf{B}'^{-1}(\mathbf{P}'_G - \mathbf{P}'_D) \\ 0 \end{bmatrix} = \mathbf{X}\begin{bmatrix} \mathbf{B}'^{-1} & 0 \\ 0 & 0 \end{bmatrix}(\mathbf{P}_G - \mathbf{P}_D) \tag{2.32}$$

Let

$$\mathbf{T} = \mathbf{X}\begin{bmatrix} \mathbf{B}'^{-1} & 0 \\ 0 & 0 \end{bmatrix} \tag{2.33}$$

The line load flow can be expressed as

$$\mathbf{F} = \mathbf{T}(\mathbf{P}_G - \mathbf{P}_D) \tag{2.34}$$

The elements in matrix $\mathbf{T}$ can be viewed as the sensitivity coefficient of branch load flow with regard to generator power output. The two kinds of direct current load flow models described above will be used many times in this book and readers should be familiar with them.

Example 2.1—For the three-node power system shown in Figure 2.3, the reactances of the three lines all equal 1. Let us employ the two methods introduced in this chapter to calculate the line flow.

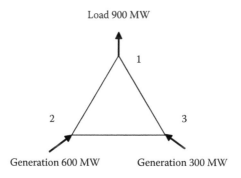

Load 900 MW

1

2   3

Generation 600 MW   Generation 300 MW

Figure 2.3   Three-node power system.

Solution—Let us try the first method for calculation. The DC load flow equation for the example power system is

$$\begin{bmatrix} P_{D1} \\ P_{G2} \\ P_{G3} \end{bmatrix} = \begin{bmatrix} 2 & -1 & -1 \\ -1 & 2 & -1 \\ -1 & -1 & 2 \end{bmatrix} \begin{bmatrix} \theta_1 \\ \theta_2 \\ \theta_3 \end{bmatrix}$$

Let node 3 serve as the reference; delete the third line and variable θ_3, and the following equation is obtained:

$$\begin{bmatrix} P_{D1} \\ P_{G2} \end{bmatrix} = \begin{bmatrix} 2 & -1 \\ -1 & 2 \end{bmatrix} \begin{bmatrix} \theta_1 \\ \theta_2 \end{bmatrix}$$

Then, solving the above equation, the nodal voltage angles are obtained:

$$\begin{bmatrix} \theta_1 \\ \theta_2 \end{bmatrix} = \begin{bmatrix} 2 & -1 \\ -1 & 2 \end{bmatrix}^{-1} \begin{bmatrix} P_{D1} \\ P_{G2} \end{bmatrix} = \begin{bmatrix} 0.6667 & 0.3333 \\ 0.3333 & 0.6667 \end{bmatrix} \begin{bmatrix} -0.9 \\ 0.6 \end{bmatrix} = \begin{bmatrix} -0.4 \\ 0.1 \end{bmatrix}$$

Notice that the load flow of each branch is

$$F_{21} = \theta_2 - \theta_1 = 0.5$$

$$F_{31} = \theta_3 - \theta_1 = 0.4$$

$$F_{23} = \theta_2 - \theta_3 = 0.1$$

Converting the above per-unit value into actual value, the load flows of each branch can be obtained. The results are 500, 400, and 300 MW. Now, let us try the second method for calculation. First, we build matrix **X** as follows:

$$\mathbf{X} = \begin{bmatrix} -1 & 1 \\ -1 & & 1 \\ & 1 & -1 \end{bmatrix}, \text{ corresponding to lines } \begin{matrix} 2-1 \\ 3-1 \\ 2-3 \end{matrix}$$

As noted above, the inverse of matrix **B′** is

$$\mathbf{B}'^{-1} = \begin{bmatrix} 0.6667 & 0.3333 \\ 0.3333 & 0.6667 \end{bmatrix}$$

Now let us calculate the matrix **T**:

$$\mathbf{T} = \mathbf{X} \begin{bmatrix} \mathbf{B}'^{-1} & 0 \\ 0 & 0 \end{bmatrix} = \begin{bmatrix} -1 & 1 & \\ -1 & & 1 \\ & 1 & -1 \end{bmatrix} \begin{bmatrix} 0.6667 & 0.3333 & 0 \\ 0.3333 & 0.6667 & 0 \\ 0 & 0 & 0 \end{bmatrix}$$

$$= \begin{bmatrix} -0.3333 & 0.3333 & 0 \\ -0.6667 & -0.3333 & 0 \\ 0.3333 & 0.6667 & 0 \end{bmatrix}$$

Finally we calculate the branch load flow:

$$\begin{bmatrix} F_{21} \\ F_{31} \\ F_{23} \end{bmatrix} = \mathbf{T}(\mathbf{P}_G - \mathbf{P}_D) = \begin{bmatrix} -0.3333 & 0.3333 & 0 \\ -0.6667 & -0.3333 & 0 \\ 0.3333 & 0.6667 & 0 \end{bmatrix} \begin{bmatrix} -0.9 \\ 0.6 \\ 0.3 \end{bmatrix} = \begin{bmatrix} 0.5 \\ 0.4 \\ 0.1 \end{bmatrix}$$

Obviously, the calculation results of the second method are the same as those from the first method.

2.3 Load Flow under Outages

When a contingency occurs in a power system, some of the generators or the transmission lines may be tripped. Let us look at the impacts of contingencies on power system dispatching. Suppose the dispatch scheme shown in Figure 2.4 is optimal. When one of the transmission lines is tripped, the load flow of the system is as

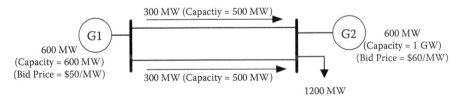

Figure 2.4 Optimal dispatch without considering the N-1 contingency.

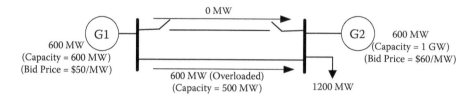

Figure 2.5   System state after contingency.

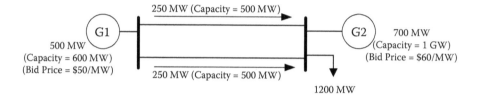

Figure 2.6   Security-constrained economic dispatch.

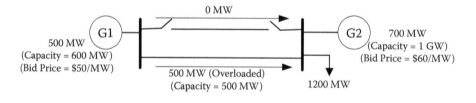

Figure 2.7   Post-contingency system state under security-constrained economic dispatch.

shown in Figure 2.5. We can see that the remaining transmission line will be overloaded, which is certainly not acceptable. The security-constrained economic dispatch is shown in Figure 2.6. Under the security-constrained dispatch scheme, if one transmission line is tripped, the system remains secure, as shown in Figure 2.7. This section is an exposition of the very basic ideas of the security-constrained dispatch—an important concept in power system operations.

2.4 Fundamentals of Constrained Optimization

Let us now look at the optimality condition of a single-variable inequality-constrained optimization problem:

$$Min \quad c(x) \tag{2.35}$$

$$S.T. \quad h(x) \le b \tag{2.36}$$

Suppose we found an optimal solution x. The inequality constraint $h(x) \leq b$ is either active ($h(x) = b$) or not active ($h(x) < b$). If the inequality constraint is not active, the optimal solution satisfies

$$\frac{dc}{dx} = 0 \tag{2.37}$$

If the inequality constraint is active, the inequality-constrained optimization problem is converted to an equality-constrained problem:

$$Min \quad c(x) \tag{2.38}$$

$$S.T. \quad h(x) = b \tag{2.39}$$

For the solution of the above problem, let us form its Lagrange function $\Gamma = c(x) + \mu[h(x) - b]$, and calculate its partial derivatives to x and μ. We obtain the following optimality conditions:

$$\frac{dc}{dx} + \mu \frac{dh}{dx} = 0 \tag{2.40}$$

$$h(x) = b \tag{2.41}$$

In summary, the optimal condition of the single-variable inequality-constrained optimization problems is

$$h(x) < b, \frac{dc}{dx} = 0 \tag{2.42}$$

or

$$h(x) = b, \frac{dc}{dx} + \mu \frac{dh}{dx} = 0 \tag{2.43}$$

Now we can examine the properties of the multiplier μ related to the inequality constraint when the constraint is active. First, if b is regarded as a parameter, the optimal solution x and the corresponding objective function value $c(x)$ will vary as b changes. Furthermore, because the size of solution set $\{x: h(x) \leq b\}$ depends on the value of b, the greater the value of b, the smaller the optimal solution of objective function $c(x)$ will be or it will remain the same. In other words, the objective function value at the optimal solution is a non-increasing function of parameter b. The optimal solution x and the corresponding Lagrangian function μ satisfy the following simultaneous equations:

$$\begin{cases} \dfrac{dc}{dx} + \mu \dfrac{dh}{dx} = 0 \\ h(x) = b \end{cases} \qquad (2.44)$$

Therefore, the optimal solution is an implicit function of b, i.e., $x = x(b)$. Substituting it into the above simultaneous equations, we get

$$h[x(b)] = b \qquad (2.45)$$

Differentiating with respect to b, we obtain

$$\frac{dh}{dx}\frac{dx}{db} = 1.$$

Substituting the formula into the optimal condition, we get

$$\frac{dc}{dx} + \mu \frac{db}{dx} = 0$$

which leads to the following important relationship:

$$\mu = -\frac{dc}{db} \qquad (2.46)$$

As stated earlier, the greater the value of b, the smaller the objective function $c(x)$ at the optimal solution. In other words, if $db > 0$, then $dc < 0$; conversely, if $db < 0$, then $dc > 0$. Therefore $\mu > 0$. From another viewpoint, the value of objective function at the optimal solution is a decreasing function of parameter b. Therefore

$$\frac{dc}{db} < 0$$

which reduces to $\mu > 0$. In other words, the Lagrangian multiplier corresponding to the inequality constraint must be positive.

To unify the optimality conditions, let the Lagrangian multiplier $\mu = 0$ when the inequality constraint is not active. We thus obtain a Lagrangian function that has a unified form $\Gamma = c(x) + \mu[h(x) - b]$. In this way, when building Lagrangian functions, we will not worry about whether the inequality is active. Therefore we will not deal with different constraints (equality or inequality) explicitly. From now on, we will build Lagrangian functions using this method. Now we can obtain a concise form of the optimal condition for single-variable inequality-constrained optimization problems:

$$
\begin{cases}
h(x) \le b \\[2mm]
\dfrac{dc}{dx} + \mu \dfrac{dh}{dx} = 0 \\[2mm]
\mu[h(x) - b] = 0 \\[2mm]
\mu \ge 0
\end{cases}
\tag{2.47}
$$

The formula $\mu[h(x) - b] = 0$ is called the complementarity relaxation condition, which requires that either μ or $[h(x) - b]$ be 0.

Based on the above analysis, we can easily calculate the optimal condition for multi-variable general inequality-constrained optimization problems. First, we define the multi-variable general inequality-constrained optimization problem:

$$
Min \quad c(\mathbf{x}) \tag{2.48}
$$

$$
S.T. \quad \mathbf{h}(\mathbf{x}) \le \mathbf{b} \tag{2.49}
$$

$$
\mathbf{g}(\mathbf{x}) = \mathbf{d} \tag{2.50}
$$

Let the Lagrangian function be

$$
\Gamma = c(\mathbf{x}) + \boldsymbol{\mu}^T[\mathbf{h}(\mathbf{x}) - \mathbf{b}] + \boldsymbol{\lambda}^T[\mathbf{g}(\mathbf{x}) - \mathbf{d}] \tag{2.51}
$$

The optimal condition in Equation 2.52 is called the Kuhn–Tucker condition:

$$
\begin{cases}
\dfrac{\partial c}{\partial \mathbf{x}} + \left(\dfrac{\partial \mathbf{h}}{\partial \mathbf{x}}\right)^T \boldsymbol{\mu} + \left(\dfrac{\partial \mathbf{g}}{\partial \mathbf{x}}\right)^T \boldsymbol{\lambda} = 0 \\[2mm]
\mathbf{h}(\mathbf{x}) \le \mathbf{b} \\[2mm]
\mathbf{g}(\mathbf{x}) = \mathbf{d} \\[2mm]
\boldsymbol{\mu}^T[\mathbf{h}(\mathbf{x}) - \mathbf{b}] = 0 \\[2mm]
\boldsymbol{\mu} \ge 0
\end{cases}
\tag{2.52}
$$

There is an additional condition that an optimal solution must satisfy, that is, at the optimal solution, the $\partial \mathbf{h}/\partial \mathbf{x}$ and $\partial \mathbf{g}/\partial \mathbf{x}$ matrix is of full low ranks. The row vector of $\partial \mathbf{h}/\partial \mathbf{x}$ is linearly independent, and the row vector of $\partial \mathbf{g}/\partial \mathbf{x}$ is linearly independent. This condition is known as the constraint qualification. Some may worry whether the constraint qualification is satisfied. The appendix of this chapter demonstrates an affirmative result.

The constraint qualification condition is almost always satisfied. We should point out that not all optimization problems have optimal solutions. One condition for a solution (Bazaraa et al. 1993) is that the objective is continuous and the set of feasible solutions is bounded and closed.

Let us now consider an example of a solution of the optimization problem using the Kuhn–Tucker condition.

Example 2.2—Solve the following inequality-constrained optimization problem:

$$\underset{y_1, y_2}{Min} \quad c(y_1, y_2) = y_1^2 + y_2^2$$

$$S.T. \quad y_1 + 2y_2 \leq 2$$

$$-y_1 - y_2 \leq -1$$

Solution—First build the Lagrangian function

$$\Gamma = y_1^2 + y_2^2 + \lambda_1(y_1 + 2y_2 - 2) + \lambda_2(1 - y_1 - y_2)$$

The optimal condition is:

$$
\begin{cases}
y_1 + 2y_2 \leq 2 \\
-y_1 - y_2 \leq -1 \\
\\
2y_1 + \lambda_1 - \lambda_2 = 0 \\
2y_2 + 2\lambda_1 - \lambda_2 = 0 \\
\\
\lambda_1(y_1 + 2y_2 - 2) = 0 \\
\lambda_2(1 - y_1 - y_2) = 0 \\
\\
\lambda_1, \lambda_2 \geq 0
\end{cases}
$$

This optimal condition contains four variables but only two linear equations. The other two equations can be found in the complementarity relaxation condition. Therefore, we can obtain four conditions:

1. $(\lambda_1, \lambda_2 = 0)$
2. $(\lambda_1, \lambda_2 \neq 0)$
3. $(\lambda_1 = 0, \lambda_2 \neq 0)$
4. $(\lambda_1 \neq 0, \lambda_2 = 0)$

Accordingly, we obtain four sets of simultaneous equations:

$$
\begin{cases}
2y_1 + \lambda_1 - \lambda_2 = 0 \\
2y_2 + 2\lambda_1 - \lambda_2 = 0 \\
\lambda_1 = 0 \\
\lambda_2 = 0
\end{cases}
\qquad
\begin{cases}
2y_1 + \lambda_1 - \lambda_2 = 0 \\
2y_2 + 2\lambda_1 - \lambda_2 = 0 \\
y_1 + 2y_2 - 2 = 0 \\
1 - y_1 - y_2 = 0
\end{cases}
$$

$$
\begin{cases}
2y_1 + \lambda_1 - \lambda_2 = 0 \\
2y_2 + 2\lambda_1 - \lambda_2 = 0 \\
\lambda_1 = 0 \\
1 - y_1 - y_2 = 0
\end{cases}
\qquad
\begin{cases}
2y_1 + \lambda_1 - \lambda_2 = 0 \\
2y_2 + 2\lambda_1 - \lambda_2 = 0 \\
y_1 + 2y_2 - 2 = 0 \\
\lambda_2 = 0
\end{cases}
$$

After solving the four sets of equations, we discard the solutions that do not satisfy the optimality condition to reach the final optimal solution.

As this analysis indicates, we suppose we have found the active constraints and accordingly the optimal condition. However, this hypothesis cannot provide any aid for solving practical problems. For linear optimization problems, a simplex algorithm can be employed for finding active constraints. Now let us introduce a graphical solution method for linear programming. Suppose the problem is

$$
\underset{Z_1, Z_2}{Max} \quad 99Z_1 + 11Z_2
$$

$$
S.T. \quad
\begin{bmatrix}
0.6667 & 0.3333 \\
0.3333 & 0.6667 \\
0.3333 & -0.3333
\end{bmatrix}
\begin{bmatrix}
Z_1 \\
Z_2
\end{bmatrix}
\leq
\begin{bmatrix}
500 \\
99999999 \\
99999999
\end{bmatrix}
$$

$$
0 \leq Z_1 \leq 900
$$

$$
0 \leq Z_2 \leq 100
$$

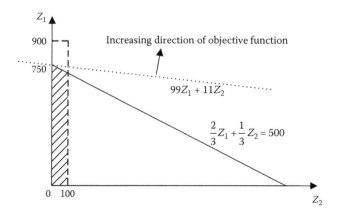

Figure 2.8 Graphical solution to linear programming.

The graphical solution for this problem is shown in Figure 2.8. For non-linear optimization problems, no generally accepted mature algorithm has been developed yet, and many algorithms are effective only for certain kinds of optimization problems. The solution method for the example problem involves enumeration. Obviously, an enumeration method is not practical because the number of equations for solving a problem will grow exponentially with the size of the problem (numbers of constraints and variables).

2.5 Security-Constrained Economic Dispatch

By combining the economic dispatch model with the DC load flow model, we can obtain the load flow-constrained economic dispatch model:

$$\underset{P_G,\theta}{Min} \sum_{i=1}^{NG} c_i(P_{Gi}) \tag{2.53}$$

$$S.T. \quad \mathbf{P}_G - \mathbf{P}_D = \mathbf{B\theta} \tag{2.54}$$

$$\mathbf{X\theta} \le \overline{\mathbf{F}} \tag{2.55}$$

$$\underline{\mathbf{P}}_G \le \mathbf{P}_G \le \overline{\mathbf{P}}_G \tag{2.56}$$

Note that we should add a constraint in the above optimization model, i.e., $-\mathbf{X\theta} \le \overline{\mathbf{F}}$, when the problem of inverse load flow will be considered. However, for ease of explanation, we have not included the constraint.

The above security-constrained economic dispatch models have been employed for practical power systems and may be expanded easily to cope with various engineering constraints. For example, when the voltage constraint is considered, we can expand the model to build an optimal power flow model. Let us review a simple example.

Example 2.3—Build up the security-constrained economic dispatch problem of the three-node power system as shown in the above section.

Solution—

$$\underset{P_G,\theta}{Min} \quad c_2(P_{G2}) + c_3(P_{G3})$$

$$S.T. \quad P_{G2} + P_{G3} - P_{D1} = 0$$

$$\begin{bmatrix} P_{D1} \\ P_{G2} \\ P_{G3} \end{bmatrix} = \begin{bmatrix} 2 & -1 & -1 \\ -1 & 2 & -1 \\ -1 & -1 & 2 \end{bmatrix} \begin{bmatrix} \theta_1 \\ \theta_2 \\ \theta_3 \end{bmatrix}$$

$$\theta_1 - \theta_2 \le \bar{F}_{12} \qquad \theta_2 - \theta_1 \le \bar{F}_{12} \qquad \theta_1 - \theta_3 \le \bar{F}_{13}$$

$$\theta_3 - \theta_1 \le \bar{F}_{13} \qquad \theta_2 - \theta_3 \le \bar{F}_{23} \qquad \theta_3 - \theta_2 \le \bar{F}_{23}$$

$$\underline{P}_{G2} \le P_{G2} \le \bar{P}_{G2}$$

$$\underline{P}_{G3} \le P_{G3} \le \bar{P}_{G3}$$

Note that $\theta_3 = 0$ is constant. Generally, this economic dispatch model in sparse form is used frequently by dispatch centers in energy management systems. For conducting pricing analysis, a dense economic dispatch model is often more convenient. After considering load flow constraints of transmission lines, the economic dispatch model in dense matrix form can be formulated as

$$\underset{P_G}{Min} \quad \sum_{i=1}^{NG} c_i(P_{Gi}) \qquad (2.57)$$

$$S.T. \quad \mathbf{e}^T(\mathbf{P}_G - \mathbf{P}_D) = 0 \qquad (2.58)$$

$$\mathbf{T}(\mathbf{P}_G - \mathbf{P}_D) \le \bar{\mathbf{F}} \qquad (2.59)$$

$$\underline{\mathbf{P}}_G \le \mathbf{P}_G \le \bar{\mathbf{P}}_G \qquad (2.60)$$

Example 2.4—Build a security-constrained economic dispatch model of the three-node system shown in the above section.

Solution—

$$\underset{P_G}{Min} \quad c_2(P_{G2}) + c_3(P_{G3})$$

$$S.T. \quad P_{G2} + P_{G3} - 0.9 = 0$$

$$\begin{bmatrix} -0.3333 & 0.3333 & 0 \\ -0.6667 & -0.3333 & 0 \\ 0.3333 & 0.6667 & 0 \end{bmatrix} \begin{bmatrix} -0.9 \\ P_{G2} \\ P_{G3} \end{bmatrix} \le \begin{bmatrix} \bar{F}_{21} \\ \bar{F}_{31} \\ \bar{F}_{23} \end{bmatrix}$$

$$\underline{P}_{G2} \le P_{G2} \le \bar{P}_{G2}$$

$$\underline{P}_{G3} \le P_{G3} \le \bar{P}_{G3}$$

The economic dispatch model introduced in this section will be used many times throughout subsequent chapters of this book. We now look at a method for solving the above problem.

Suppose we have found the active constraints. Let $\mathbf{T}_\Omega$ and $\bar{\mathbf{F}}_\Omega$ denote $\mathbf{T}$ and $\bar{\mathbf{F}}$ to correspond to the active constraints accordingly; Ω denotes the active constraint set of minimum power output limits, and $\widehat{\Omega}$ denotes the active constraint set of maximum power output limits. Therefore, we are faced with an equality-constrained optimization problem as follows:

$$\underset{P_G}{Min} \quad \sum_{i=1}^{NG} c_i(P_{Gi}) \tag{2.61}$$

$$S.T. \quad \mathbf{e}^T(\mathbf{P}_G - \mathbf{P}_D) = 0 \tag{2.62}$$

$$\mathbf{T}_\Omega(\mathbf{P}_G - \mathbf{P}_D) = \bar{\mathbf{F}}_\Omega \tag{2.63}$$

$$P_{Gi} = \bar{P}_{Gi}, i \in \widehat{\Omega} \tag{2.64}$$

$$P_{Gj} = \underline{P}_{Gj}, j \in \widecheck{\Omega} \tag{2.65}$$

Similarly, we define the Lagrangian function:

$$\Gamma = \sum_{i=1}^{NG} c_i(P_{Gi}) + \lambda \mathbf{e}^T (\mathbf{P}_G - \mathbf{P}_D) + \mu^T [\mathbf{T}_\Omega (\mathbf{P}_G - \mathbf{P}_D) - \overline{\mathbf{F}}_\Omega] +$$

$$\sum_{i \in \Omega} \hat{\tau}_i (P_{Gi} - \overline{P}_{Gi}) - \sum_{j \in \Omega} \check{\tau}_j [(P_{Gj} - \underline{P}_{Gj})] \tag{2.66}$$

Differentiating the above Lagrangian function, we obtain the following set of extreme point equations:

$$\frac{\partial \Gamma}{\partial P_{Gi}} = \frac{\partial c_i}{\partial P_{Gi}} + \lambda + \sum_k \mu_k T_{ki} + \hat{\tau}_i - \check{\tau}_i = 0 \tag{2.67}$$

$$\frac{\partial \Gamma}{\partial \mu} = \mathbf{T}_\Omega (\mathbf{P}_G - \mathbf{P}_D) - \overline{\mathbf{F}}_\Omega = 0 \tag{2.68}$$

$$\frac{\partial \Gamma}{\partial \hat{\tau}_i} = P_{Gi} - \overline{P}_{Gi} = 0, \, i \in \widehat{\Omega} \tag{2.69}$$

$$\frac{\partial \Gamma}{\partial \check{\tau}_j} = P_{Gj} - \underline{P}_{Gj} = 0, \, j \in \check{\Omega} \tag{2.70}$$

For ease of illustration, we define the Lagrangian multiplier for inactive constraints as 0. Therefore, the optimal solution related to inequality constraints can be formulated as the following complementary relaxation form (or condition):

$$\begin{cases} \mu^T [\mathbf{T}(\mathbf{P}_G - \mathbf{P}_D) - \overline{\mathbf{F}}] = 0 \\ \check{\tau}^T (\underline{\mathbf{P}}_G - \mathbf{P}_G) = 0 \\ \hat{\tau}^T (\mathbf{P}_G - \overline{\mathbf{P}}_G) = 0 \end{cases} \tag{2.71}$$

The optimal (Kuhn–Tucker) condition can be summarized as

$$\left\{ \begin{array}{l} \mathbf{e}^T(\mathbf{P}_G - \mathbf{P}_D) = 0 \\[2mm] \mathbf{T}(\mathbf{P}_G - \mathbf{P}_D) \leq \overline{\mathbf{F}} \\[2mm] \underline{\mathbf{P}}_G \leq \mathbf{P}_G \leq \overline{\mathbf{P}}_G \\[4mm] \dfrac{\partial c_i}{\partial P_{Gi}} + \lambda + \mu_k \displaystyle\sum_k T_{ki} + \hat{\tau}_i - \check{\tau}_i = 0, \quad i = 1, 2, \ldots, NG \\[4mm] \boldsymbol{\mu}^T[\mathbf{T}(\mathbf{P}_G - \mathbf{P}_D) - \overline{\mathbf{F}}] = 0 \\[2mm] \check{\boldsymbol{\tau}}^T(\underline{\mathbf{P}}_G - \mathbf{P}_G) = 0 \\[2mm] \hat{\boldsymbol{\tau}}^T(\mathbf{P}_G - \overline{\mathbf{P}}_G) = 0 \\[4mm] \boldsymbol{\mu}, \hat{\boldsymbol{\tau}}, \check{\boldsymbol{\tau}} \geq 0 \end{array} \right. \tag{2.72}$$

Let NL denote the number of transmission lines. Based on $1 + NL + 3 * NG$ equations, which equals the number of variables, the above system is well defined. In the Kuhn–Tucker condition, equality and inequality constraints are processed in a unified way. Similarly, when building a Lagrangian function, the equality and inequality constraints can be processed as follows:

$$\Gamma = \sum_{i=1}^{NG} c_i(P_{Gi}) + \lambda \mathbf{e}^T(\mathbf{P}_G - \mathbf{P}_D) + \boldsymbol{\mu}^T[\mathbf{T}(\mathbf{P}_G - \mathbf{P}_D) - \overline{\mathbf{F}}] +$$
$$\hat{\boldsymbol{\tau}}^T(\mathbf{P}_G - \overline{\mathbf{P}}_G) + \check{\boldsymbol{\tau}}^T(\underline{\mathbf{P}}_G - \mathbf{P}_G) \tag{2.73}$$

We will adopt the above general method to set up Lagrangian functions. It is important to note that the production-grade calculation process of security-constrained economic dispatch may be far more complicated than presented in this section. As shown in Figure 2.9, the calculation procedure includes automatic contingency selection, contingency evaluation, identification of active constraints, and solving of linear programming models.

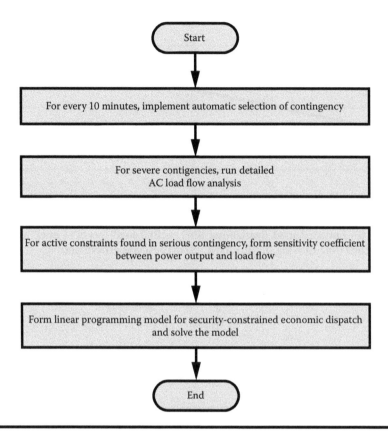

Figure 2.9 Calculation process for solving security-constrained economic dispatch.

2.6 Load Frequency Control

A modern power system has tens of thousands of customers. The system load varies from time to time because customers constantly turn electrical equipment (lighting, computers, appliances) on and off. If a utility's generation equipment cannot regulate its power output in a real-time fashion to follow varying load demands, generator rotor speeds will change abnormally and cause the system frequency to exceed its prescribed limit. This must be prevented during normal operation because frequency deviations will damage generation equipment.

Furthermore, certain loads such as electric motors cannot bear large frequency deviations. For these reasons, the generation equipment in a power system must be equipped with control mechanisms that can quickly regulate power output to follow the variations of loads and thus maintain system frequency in a certain range. This is the basic concept of frequency adjustment. Figure 2.10 shows a five-minute system load and the corresponding generator outputs of the Electric Reliability Council of Texas (ERCOT) power system in the United States.

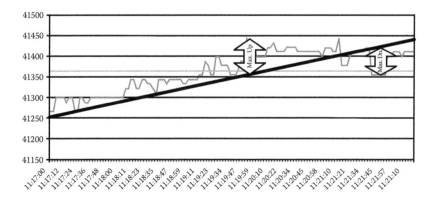

Figure 2.10   Forecast load (horizontal line), actual load (vertical line), and generator output (line moving up in x-axis). (Source: www.eroct.com)

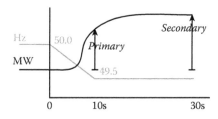

Figure 2.11   Primary and secondary frequency adjustments.

Two kinds of methods are generally employed for frequency regulation. The first is use of a turbine governor to follow load variations. This method is called primary frequency control. It provides fast response and is very effective for frequency control after the tripping of large generators within a few seconds. The drawback is the presence of frequency deviation after the system settles down. In other words, this method cannot restore the system frequency to a nominal value (for example, 50 Hz). The secondary control measure is to remove this frequency deviation, as shown in Figure 2.11.

Modern power systems are generally controlled in a decentralized way. Regional subsystems are interconnected by transmission lines to exchange electric power (Figure 2.12). For purposes of electricity energy settlements among regional subsystems, control of interconnection line flows is required. This interconnected flow control system is integrated into the frequency control system shown. The combination of secondary frequency control and interconnection line flow control is called automatic generation control (AGC).

Not all generators are equipped with (secondary) automatic generation control systems. Generally, as shown in Figure 2.13, generators equipped with AGC systems

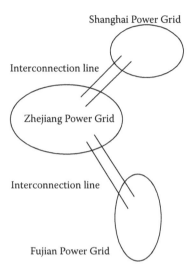

Figure 2.12   Interconnection of power grids.

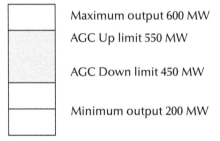

Figure 2.13   Domain of power output for AGC.

can utilize the control only under certain power output conditions. Furthermore, the regulation speed of AGC is limited.

We encounter a subtlety when modeling the automatic generation capacity (or reserve) of a unit, that is, we must differentiate between positive capacity A^+ and negative capacity A^-. Let v_A be the output that a generator can increase or decrease in 1 minute according to AGC signals. Suppose a dispatching period is 10 minutes. The generator output that can be regulated in 10 minutes is less than 10 v_A.

Generally, a generator can provide AGC service only in a certain domain of its output range. If the actual output of a unit is out of the domain, the unit cannot function well (this relates to fuel burning stability). Let $\bar{A}$ and $\underline{A}$ denote the up and down limits of the domain of capacity—designated the AGC up limit and AGC down limit, respectively. The following constraint for automatic generation capacity of a generator should be satisfied:

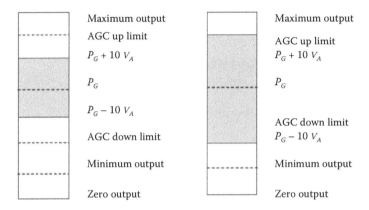

Figure 2.14 Regulation capacities of two generators.

$$A^+ \leq \min\left\{\bar{A} - P_G, \; 10v_A\right\} \qquad (2.74)$$

$$A^- \leq \min\left\{P_G - \underline{A}, \; 10v_A\right\} \qquad (2.75)$$

Figure 2.14 depicts the regulation capacities of two generators. The generator at the left has a slow regulation speed and the one at the right has a fast regulation speed. The AGC service that a generator can provide in a dispatch period (for example, 10 minutes) is called the regulation capacity of the dispatch period. Therefore, the regulation capacity of the generator at the left is 10 v_A (regulation capacity is constrained by speed). The regulation capacity of the generator at the right is equal to the up limit of regulation capacity power output P_G (regulation capacity is constrained by the up limit). The regulation speed is slow when v_A is small and fast when v_A is large.

All modern power systems are equipped with adequate AGC capacity. When loads are subject to fast variations, for example between 6 and 8 a.m., more AGC capacity should be assigned. Meanwhile, when load variations are smaller, less AGC capacity can be assigned. If many generators can provide automatic control service in an electricity market, the problem needs to be handled during the design phase by determining which generator should be selected to finish the task. The AGC generators may be selected through auctions. Later in this chapter, we will introduce an auction model. A system's demand for AGC capacity is an operation constraint that may be formulated as an inequality constraint as follows:

$$\sum_i A_i^+ \geq D_A \qquad (2.76)$$

$$\sum_i A_i^- \geq D_A \qquad (2.77)$$

where, D_A is the system demand for generation control capacity determined by a statistical study of daily load curves. For example, when loads vary quickly during mornings, a large D_A should be selected. However, when the load variation is slower at other times, a small D_A should be selected to reduce the wear and tear on generators.

Costs will be incurred for providing AGC services. The generators that provide AGC service should be compensated. Otherwise, the generators will lose incentives for providing AGC service. Analysis of the cost helps formulate AGC service market rules.

The AGC service cost includes fixed costs (purchases of equipment) and variable costs. Hirst and Kirby (1996) provided a detailed account of AGC cost auditing. The fixed costs are investments for (1) communication equipment and (2) control equipment and supervision system upgrades. Variable costs include (1) equipment maintenance and operation costs, (2) depreciation (frequent output regulation and adjustment increase depreciation costs for primary equipment) and (3) increased fuel consumption caused by fluctuations of generator outputs.

2.7 Spinning Reserve

Equipment made by humans may break down and power generation equipment is no exception. System frequency will decrease if a generator is tripped. In an effort to maintain system frequency, the outputs of the other generators should be increased to satisfy the load. A system must be equipped with enough fast response generators, such as hydroelectric and thermal units, because the frequency cannot be kept at a low level for a long time. The fast response generators can increase their power output quickly to satisfy load demand after the tripping of one generator and are sometimes called reserve generators. A power increase within 10 minutes is called 10-minute reserve capacity.

Generally speaking, a deliberately designed combination of reserves is needed for a modern power system (for example, 10-second, 60-second, 10-minute, and 30-minute reserves). However, in this book, only 10-minute reserves are discussed. The relationship of the reserve capacity of a generator with its maximum available capacity is shown in Figure 2.15. Based on the figure, we can see that the reserve capacity of a generator should satisfy the following inequality constraint:

$$R \leq \min\left\{\overline{P_G} - P_G, \ 10^* v_R\right\} \tag{2.78}$$

where, v_R denotes the outputs of generators that can respond in 1 minute according to the instructions of the plant operator. The difference between v_R and v_A is that v_A denotes the output that can be increased in 1 minute according to AGC signals. One common operational requirement is that the sum of generators' 10-minute

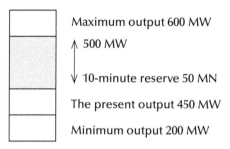

Figure 2.15   Reserve capacity versus maximum available capacity.

reserves should be greater than a specified value. This requirement can be formulated as

$$\sum_i R_i \geq D_R \tag{2.79}$$

where D_R is the reserve requirement often set at 10% of system demand or the capacity of the largest generator online. If many generators can provide reserve service in an electricity market, they can be selected through auction. In the next chapter, we will introduce an optimal dispatch model for reserve capacity. The issue of opportunity cost associated with providing reserves will be covered in detail later in this book.

2.8 Generation Scheduling

Wear-and-tear and extra fuel costs are incurred during start-ups of thermal units. Similarly, wear-and-tear costs are incurred for starting hydroelectric units. Although unit start-up costs vary, they must be considered when doing generation scheduling and designing markets. Hydroelectric units can increase their power outputs to available capacities in 1 minute or less, but thermal and nuclear generators generally require more time for start-up.

Minimum operating hours, minimum power output, and other factors should be considered when scheduling thermal and nuclear generators. Because system loads vary continuously, the daily operation of a power system should include proper scheduling of start-ups and shut-downs of generators to meet load demands. The schedule of a three-unit system is shown in Figure 2.16.

A generator can be either on-line or off-line. Therefore, the generator scheduling program can be viewed as a 0–1 integer program. As noted earlier, additional fuel costs will be incurred for start-ups of thermal units, and wear-and-tear costs will arise from start-ups of hydroelectric units. Therefore, both fuel and wear costs should

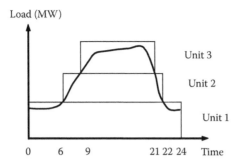

Figure 2.16 Scheduling a three-unit system.

be considered when calculating generation costs. As with a security-constrained economic dispatch problem, load flow constraints should also be considered when devising daily generation schedules.

Naturally, a minimum total cost generation schedule can be formulated as a mathematical problem. Let $U_i^t = 1$ indicate that generator i is on at hour t, $U_i^t = 0$ indicate that generator i is off at hour t, $\mathbf{U}_i$ denote a vector composed of a variable U_i^t, $\mathbf{U}_i = [U_i^1, U_i^2, ..., U_i^{24}]^T$, $c_i(P_{Gi}^t)$ denote the fuel cost of generator i, and the superscript t denote time in a day. Generally, the start-up cost is a function of a unit commitment decision variable that may be denoted as $S_i^t(U_i)$. Quite often $S_i^t(U_i)$ is an exponential function of shut-down time:

$$S^t = a + b\left[1 - \exp\left(-\frac{shutdown\_time}{c}\right)\right] \tag{2.80}$$

In an electricity market environment, $S_i^t(U_i)$ can be constant, i.e., it can represent the start-up price of a generator. The start-up cost of a generator exceeds 0 at the moment of start-up. At other moments, the start-up cost is 0. Therefore, in the objective function, the cost can be formulated as $S_i^t(U_i)U_i^t(1 - U_i^{t-1})$.

This formula accurately demonstrates that the start-up cost of a generator is greater than 0 only at the moment of start-up, because the formula is not equal to 0 only when $U_i^t = 1$ and $U_i^{t-1} = 0$. Therefore, the objective function of the generation scheduling problem is composed of two parts, formulated as follows:

$$\underset{P_G, U}{Min} \sum_{t=1}^{24} \sum_{i=1}^{NG} \left[U_i^t c_i(P_{Gi}^t) + S_i^t(\mathbf{U}_i)U_i^t(1 - U_i^{t-1})\right] \tag{2.81}$$

Load generation balance and transmission line load flow constraints must be satisfied constantly during power system operation. These constraints can be summarized as

$$\mathbf{e}^T(\mathbf{P}_G^t - \mathbf{P}_D^t) = 0, \; \mathbf{T}(\mathbf{P}_G^t - \mathbf{P}_D^t) \leq \overline{\mathbf{F}}, \; (t = 1, 2, \ldots, 24) \qquad (2.82)$$

For ease of explanation, generator ramping rates are not considered here. The reserve constraints can be expressed as

$$\sum_{i=1}^{NG} U_i^t \overline{P}_i \geq \mathbf{e}^T \mathbf{P}_D^t + D_R, \; (t = 1, 2, \ldots, 24) \qquad (2.83)$$

Obviously, capacity constraint should also be satisfied for generator power output. The power output of an off-line generator must be 0. The constraint for generator power output can be expressed as

$$U_i^t \underline{P}_{Gi}^t \leq P_{Gi}^t \leq U_i^t \overline{P}_{Gi}^t, \; (t = 1, 2, \ldots, 24) \; (i = 1, 2, \ldots, NG) \qquad (2.84)$$

We can now summarize the aforementioned models (including the objective function and constraints). The complete mathematical model for generation scheduling can be formulated as

$$\underset{\mathbf{P}_G, \mathbf{U}}{Min} \quad \sum_{i=1}^{NG} \sum_{t=1}^{24} \left[U_i^t c_i(P_{Gi}^t) + S_i^t(\mathbf{U}_i) U_i^t(1 - U_i^{t-1}) \right] \qquad (2.85)$$

$$S.T. \quad \mathbf{e}^T(\mathbf{P}_G^t - \mathbf{P}_D^t) = 0, \; (t = 1, 2, \ldots, 24) \qquad (2.86)$$

$$\mathbf{T}(\mathbf{P}_G^t - \mathbf{P}_D^t) \leq \overline{\mathbf{F}}, \; (t = 1, 2, \ldots, 24) \qquad (2.87)$$

$$\sum_{i=1}^{NG} U_i^t \overline{P}_i \geq \mathbf{e}^T \mathbf{P}_D^t + D_R, \; (t = 1, 2, \ldots, 24) \qquad (2.88)$$

$$U_i^t \underline{P}_{Gi}^t \leq P_{Gi}^t \leq U_i^t \overline{P}_{Gi}^t, \; (t = 1, 2, \ldots, 24) \; (i = 1, 2, \ldots, NG) \qquad (2.89)$$

$$U_i^t \in \{0,1\}, \; (t = 1, 2, \ldots, 24) \; (i = 1, 2, \ldots, NG) \qquad (2.90)$$

The mathematical model for generation scheduling in actual system operation is more complex. For example, the generator ramping constraint can be formulated as

$$-\Delta\bar{\mathbf{P}}_G \leq \mathbf{P}_G^t - \mathbf{P}_G^{t-1} \leq \Delta\bar{\mathbf{P}}_G \qquad (2.91)$$

where $\Delta\bar{\mathbf{P}}_G$ is the maximum ramping power for a time period. Such constraints can be added to the above generation scheduling model easily if necessary. Obviously, if the decisions of generation scheduling are obtained via enumeration, the problem is reduced to a series of security-constrained economic dispatch problems. We can find an optimal solution for each such individual problem and then the overall problem is solved. This method is not practical because it requires exceedingly large calculation times. More efficient methods include branch-and-bound, dynamic programming, Lagrange relaxation, and others. This is an active research area that we will not pursue here.

2.9 Calculation of Transfer Capabilities of Transmission Interfaces

The transfer capability concept is widely used in on-line dispatch and for system planning. Often the physical limits of transmission systems can be approximately represented by interface transfer limits (Hamoud 2000). Transfer capabilities are important factors in real-time trading and investment reconciliation in emerging electricity markets. For example, the transmission constraints in the New England power system are encapsulated into about 20 interface limits. A similar approach is utilized in Australia and New Zealand power systems.

2.9.1 Description of Min–Max Transfer Capability Problems

Suppose we are interested in finding the transfer capability of a transmission interface composed of lines 1–3 and 2–3, as illustrated in Figure 2.17. For simplicity, we do not consider line contingencies in this example. An optimization problem illustrated below can be formulated to compute the transfer capability:

$$\underset{\mathbf{P}}{Max} \quad P_1 + P_2 \qquad (2.92)$$

$$S.T. \quad T_{11}P_1 + T_{12}P_2 \leq \bar{F}_1 \qquad (2.93)$$

$$0 \leq P_1 \leq \bar{P}_1 \qquad (2.94)$$

$$0 \leq P_2 \leq \bar{P}_2 \qquad (2.95)$$

$$0 \leq P_3 = L - P_1 - P_2 \leq \bar{P}_3 \qquad (2.96)$$

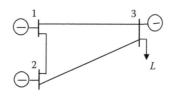

Figure 2.17   Single-line diagram of an example power system.

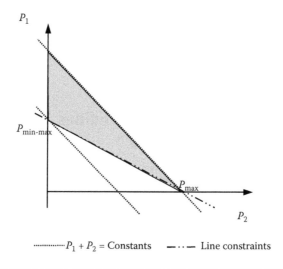

Figure 2.18   Graphical solution to the example problem.

In the above example, T denotes distribution factors, P denotes generator output, $\overline{P}$ is maximum generator capability, L is the load at bus 3. The solution of the above problem is the point P_{max} illustrated in Figure 2.18. This point yields an optimistic estimate of transfer capability because the operating points in the *shaded area* are not acceptable. It is obvious that $P_{min-max}$ is a conservative solution of transfer capability. We can see that all the operating points below the line $P_1 + P_2 = constant$ that passes $P_{min-max}$ are acceptable.

In practical applications, engineers often must find the min–max point. For example, transmission planners always want to find the maximum and min–max transfer capabilities. Maximum transfer capability indicates the upper bound that corresponds to the optimistic generation dispatch condition. Min–max transfer capability represents the lower bound of transfer capability that corresponds to the pessimistic (worst case) dispatch condition. The two numbers together indicate to planners how "strong" a system is.

Another possible application of min–max transfer capability is in new generation interconnection system impact studies. To study the impact of a proposed new generation system on the existing interface transfer capability, we always compare the new generator against a generator with similar capacity or several generators within the interface to determine whether the addition will degrade the existing interface transfer capability. However, the chosen generators to be dispatched against the new generator represent only reasonable conditions—not the worst condition.

Determining the worst dispatch condition that corresponds to the min–max point of the new generator is valuable in this kind of interconnection system impact study. Also, instead of posting maximum transfer capabilities that may not be reasonable on transmission reservation systems, posting both maximum and min–max transfer capabilities seems to be more appropriate.

Min–max transfer capability may be found using an exhaustive search of generation load space using a load flow program. However, such a search is not desired for obvious reasons. The following sections describe a method that can be used to find the exact min–max transfer capability without an exhaustive search.

From the example described above, it is clear that min–max transfer capability must satisfy two conditions: (1) the amount of power can be transferred without violating any operating constraints and (2) the ability to transfer power even if generation loads change *arbitrarily*. The second class of conditions can be restated in the following equivalent form: even under a worst generation load profile, none of the line flows exceeds the thermal limits. Mathematically, the concept is formulated as the following bi-level optimization problem:

$$\underset{P',P''}{Max} \quad \sum_{i=1}^{NLI} \sum_{j=1}^{NGen} T_{ij} \cdot (P'_j - L_j) \tag{2.97}$$

$$S.T. \quad \sum_j P'_j = \sum_j L_j \tag{2.98}$$

$$\sum_j T_{ij} \cdot (P'_j - L_j) \le \bar{F}_i \quad (i = 1, 2, \ldots, NTransC) \tag{2.99}$$

$$\underline{P}_j \le P'_j \le \bar{P}_j \quad (j = 1, 2, \ldots, NGen) \tag{2.100}$$

$$\left\{ \sum_j T_{kj} \cdot (P''^k_j - L_j) \le \bar{F}_k \right. \tag{2.101}$$

$$\underset{P''^k}{Max} \quad \sum_j T_{kj} \cdot (P''^k_j - L_j) \tag{2.102}$$

$$S.T. \quad \sum_j P_j''^k = \sum_j L_j \tag{2.103}$$

$$\underline{P}_j \le P_j''^k \le \bar{P}_j \quad (j = 1, 2, ..., NGen) \tag{2.104}$$

$$\sum_{i=1}^{NLI}\sum_{j=1}^{NGen} T_{ij} \cdot (P_j'' - L_j) \le \sum_{i=1}^{NLI}\sum_{j=1}^{NGen} T_{ij} \cdot (P_j' - L_j) \tag{2.105}$$

$$\left.\right\} \quad (k \in K)$$

where

P′ is a fictitious dispatch that results in min–max transfer capability.

P′′ᵏ is a fictitious dispatch that results in maximum line flow at line k.

NLI is the number of lines of an interface.

NGen is the number of generators.

T_{ij} is the distribution factor of line i with respect to generator j.

L_j is the active power load at the j-th node.

$\bar{F}_i$ represents the thermal limits of the i-th transmission line.

NTranC is the number of transmission constraints.

K contains the indices of line flow constraints associated with the interface under study.

Note that the above equations attempt to include both normal state and post-contingency state limits. The number of second level optimization problems is equal to the number of elements in K. For example, suppose an interface consists of four lines. The number of lower level problems, taking into account $n − 1$ contingencies, is equal to $4 + (4 * 3) = 16$. In the above formulas, loads **L** are modeled as constants, but it is straightforward to model them as variables with upper and lower bounds.

In other words, the above two-level optimization problem can be restated as follows. The optimization intends to find the transfer capability under two classes of constraints: regular operation constraints and worst-case scenario constraints in generation space. The familiar maximum transfer capability satisfies the first class of constraints only. One attractive characteristic of min–max transfer capability is that the normal and post-contingency state physical limits of each transmission line in an interface are encapsulated into a *single* constraint:

$$\sum_{j=1}^{NGen}\left(\sum_{i=1}^{NLI} T_{ij}\right) \cdot (P_j - L_j) \le \sum_{j=1}^{NGen}\left(\sum_{i=1}^{NLI} T_{ij}\right) \cdot (P_j'^* - L_j) = MMTC \tag{2.106}$$

where $\mathbf{P}'^{*}$ is the solution of the optimization problem, $\mathbf{P}$ is the generation output of the problem of interest, say, a real-time dispatch, and *MMTC* denotes min–max transfer capability. This explains why the physical limits of a transmission system can be represented by a number of interface limits.

2.9.2 Optimality Condition and Algorithm

Next we describe the optimality condition and a simple algorithm for the two-level optimization problem. For simplicity, suppose temporarily that we have only one lower-level problem. The above two-level optimization problem can be expressed in the following compact form:

$$\underset{\mathbf{P}',\mathbf{P}''}{Max} \quad \mathbf{C}'^{T} \cdot \mathbf{P}' \tag{2.107}$$

$$S.T. \quad \mathbf{A}' \cdot \mathbf{P}' + \mathbf{A}' \cdot \mathbf{P}'' \leq \mathbf{b}' \tag{2.108}$$

$$\underset{\mathbf{P}''}{Max} \quad \mathbf{C}''^{T} \cdot \mathbf{P}'' \tag{2.109}$$

$$S.T. \quad \mathbf{B}' \cdot \mathbf{P}' + \mathbf{B}'' \cdot \mathbf{P}'' \leq \mathbf{b}'' \tag{2.110}$$

The meanings of coefficients are analogous to those in the two-level optimization problem. Conceptually, we can replace the lower-level optimization problem with its Kuhn–Tucker condition and thus the bi-level optimization problem can be reformulated as a regular nonlinear program that is unfortunately highly nonconvex. A popular approach is to solve a sequence of linear programming problems defined as follows:

$$\underset{\mathbf{P}',\mathbf{P}''}{Max} \quad \mathbf{C}'^{T} \cdot \mathbf{P}' \tag{2.111}$$

$$S.T. \quad \mathbf{A}' \cdot \mathbf{P}' + \mathbf{A}' \cdot \mathbf{P}'' \leq \mathbf{b}'' \tag{2.112}$$

$$\mathbf{B}'_{b} \cdot \mathbf{P}' + \mathbf{B}''_{b} \cdot \mathbf{P}'' = \mathbf{b}''_{b} \tag{2.113}$$

where $\mathbf{B}''_{b}$ is a basis of the lower-level problem and $\mathbf{B}'_{b}$ and $\mathbf{b}''_{b}$ are the corresponding submatrices. Apparently, the solution of this simple linear program must coincide with the optimal solution of the bi-level optimization problem if the basis $\mathbf{B}'_{b}$ meets, in addition to the feasibility of the lower-level problem, the Kuhn–Tucker condition of the lower level optimization problem:

$$\mathbf{B}''^{T}_{b} \cdot \lambda''_{b} = -\mathbf{C}'' \tag{2.114}$$

$$\lambda_b'' > 0 \tag{2.115}$$

where λ_b'' contains Lagrange multipliers of the lower-level problem. When seeking an optimal solution, we can employ a branch-and-bound algorithm. In fact, the problem can be converted into a 0–1 integer program. It is not difficult to see that a branch-and-bound algorithm can extend readily to solve a problem that contains multiple second-level optimization problems.

The aforementioned algorithm yields an accurate solution. The disadvantage is that it requires exceedingly long CPU time for large-scale problems because of the famous "combinatorial explosion" phenomenon. This algorithm will be used only to benchmark a more efficient solution algorithm to be described in the next section.

2.9.3 Bi-Sectioning Search Algorithm

First, let us study the structure of the bi-level optimization problem (1–1) to (2–4). Notice that the solution of the upper-level problem (1–1) to (1–4), max transfer capability, is the upper bound of min–max transfer capability. If none of the lower-level constraints (2–0) is binding, max transfer capability coincides with min–max transfer capability. A lower bound of min–max transfer capability is 0. Let

$$x = \sum_{i=1}^{NLI} \sum_{j=1}^{NGen} T_{ij} \cdot (P_j' - L_j) \tag{2.116}$$

It follows that

$$0 \le x \le max\_transfer\_capability \tag{2.117}$$

The k-th lower-level optimization problem can be restated as follows:

$$\underset{\mathbf{P}^{''k}}{Max} \quad F_k = \sum_j T_{kj} \cdot (P_j''^k - L_j) \tag{2.118}$$

$$S.T. \quad \sum_j P_j''^k = \sum_j L_j \tag{2.119}$$

$$\underline{P}_j \le P_j''^k \le \overline{P}_j \quad (j = 1, 2, \ldots, NGen) \tag{2.120}$$

$$\sum_{i=1}^{NLI} \sum_{j=1}^{NGen} T_{ij} \cdot (P_j'' - L_j) \le x \tag{2.121}$$

The correspondence between the solution of the above problem F_k^* and x is obvious. Let us denote this correspondence as $F_k^* = F_k(x)$. Note that $F_k(x)$ has three properties.

First, it is a continuous, single-valued function of x. This property holds as long as the optimal solution of the optimization problem exists and the objective and constraint functions are continuous. Second, it is non-decreasing. In other words, as x decreases, the maximum flow on each individual line F_k^* can only decrease or remain the same. Third, the set $\Gamma_k = \{x : F_k(x) \le \bar{F}_k\}$ is convex for the following reason. Suppose x_1 and x_2 belong to Γ_k. Any point within the interval $[x_1, x_2]$ also belongs to Γ_k. Now let us reformulate the basic two-level optimization problem as the following one-dimensional optimization:

$$\underset{x}{Max} \quad x \tag{2.122}$$

$$S.T. \quad F_k(x) \le \bar{F}_k, \quad k \in K \tag{2.123}$$

$$0 \le x \le max\_transfer\_capability \tag{2.124}$$

The intersection of the convex sets Γ_k, $k \in K$, is convex. Therefore the above optimization problem has a convex feasible set. This implies that the above problem has a unique solution if the feasible set is non-empty. This problem differs from ordinary one-dimensional optimization problems in that the constraint condition is implicitly (versus explicitly) defined. The structure and solution of this optimization problem are further illustrated in Figure 2.19. *MMTC* and *MTC* indicate min–max transfer capability and maximum transfer capability, respectively.

Obviously, the optimization problem can be solved using the following informal algorithm. Let $x = MTC$ solve the problem and check the constraints for k, $k \in K$. If the constraints are satisfied, x is the solution. If not, set x to a smaller number, solve the problem, and check the constraints again. A formal bi-sectioning algorithm is illustrated in Figure 2.20.

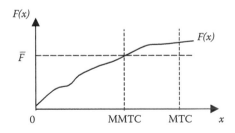

Figure 2.19 Structure and solution of the optimization problem.

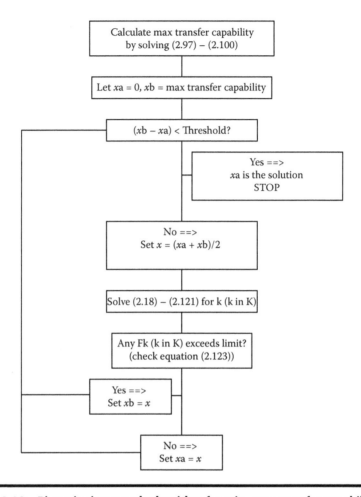

Figure 2.20 Bi-sectioning search algorithm for min–max transfer capability.

2.9.4 Difference between Min–Max and Maximum Transfer Capability

The parameters of the three-bus example system illustrated in Figure 2.17 are summarized by

$$\overline{P_1} = 200 \ , \ \overline{P_2} = 200 \ , \ \overline{P_3} = Big$$

$$\overline{F_{12}} = Big \ , \ \overline{F_{13}} = 140 \ , \ \overline{F_{23}} = Big$$

$$x_{12} = 0.01 \ , \ x_{13} = 0.01 \ , \ x_{23} = 0.02$$

The distribution factors of line 2–3 = [0.25   0.5   0.0].
    Under the above conditions, the calculation results are as follows:

Min–max transfer capability = 186.67

$$\mathbf{P}' = \begin{bmatrix} 0.0 & 186.67 & 213.33 \end{bmatrix}$$

$$\mathbf{P}'' = \begin{bmatrix} 186.67 & 0.0 & 213.33 \end{bmatrix}$$

Maximum transfer capability = 253.33

$$\mathbf{P} = \begin{bmatrix} 53.33 & 200.00 & 146.67 \end{bmatrix}$$

From the above results, we can easily verify that the dispatch of the lower-level problem is the severest combination of generation output. It is not difficult to see that the difference between min–max and maximum transfer capabilities is caused by the asymmetry of distribution factors associated with the binding constraint.

2.9.5 Conditional Min–Max Transfer Capabilities

In this and subsequent sections, we report results tested on an aggregated 118-bus system whose original single line diagram can be found at http://www.powerworld. com. The interface under study consists of the following transmission lines: 62–67, 62–66, 64–65, 59–54, 59–55, 59–56, and 59–56.

    Apparently, min–max transfer capabilities are generally conservative when used in system operation. To reduce their conservativeness, we can search for min–max transfer capabilities under different system conditions. The output of such a study is called conditional min–max transfer capability. The results related to conditional min–max transfer capabilities appear in Table 2.1. We found that the introduction of conditional min–max transfer capability reduces the conservativeness of the capability at a cost of having to solve more optimization problems.

    The impact of system conditions on min–max transfer capability is further illustrated in Figure 2.21 for the case of a three-bus system. We can see that the min–max transfer capability without a condition (point $P_{min-max}^1$) is more conservative than the min–max transfer capability with a condition (point $P_{min-max}^2$). In other words, if we know that generator 2 outputs at a minimum level, a better estimation of min–max transfer capability can be obtained in the sense that it is closer to the maximum transfer limit.

Table 2.1   Conditional Min–Max Transfer Capabilities[a]

Generator Output at Bus 66	Conditional Min–Max Transfer Capability
0	982
200	1080
400	1180
600	1283
800	1386
1000	1488
1200	1594

[a] Maximum transfer capability = 2495 MW.

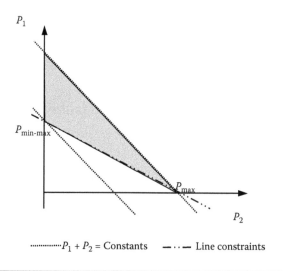

................$P_1 + P_2$ = Constants   — ·· — Line constraints

Figure 2.21   Conditional min–max transfer capability ($P^2_{min-max}$).

2.9.6 Results of Search Algorithm

Again, we compute min–max transfer capability for the interface of an IEEE 118-bus system, but this time we use the bi-sectional search algorithm described in Figure 2.20. The start point of the bi-sectioning search is the value of maximum transfer capability (2500 MW obtained from the first-level optimization). The bi-sectional algorithm yields the min–max transfer capability of 980 MW after nine iterations, while the accurate solution is 982 MW obtained from the two-level

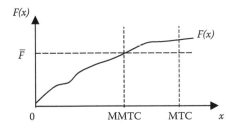

Figure 2.22 Iteration behavior of the search algorithm.

optimization (first row in Table 2.1). This demonstrates the feasibility and accuracy of the proposed bi-sectioning search algorithm for min–max transfer capability calculation. The iteration behavior is further illustrated in Figure 2.22.

2.10 Overview of Power System Operation

The main components of power systems such as power plants and substations are interconnected by a transmission network, as shown in Figure 2.23. As stated earlier, a transmission network is often interconnected with other (external) transmission networks. A power dispatch center maintains communication links with power plants and substations and is responsible for the operation and coordination of the components of the system. The tasks of a power dispatch center can be classified chronologically as follows:

- Year(s) ahead: perform long-term load forecast, generation and transmission reinforcement planning.
- Week and month ahead: calculate long-term operation parameters such as maximum transfer capabilities of transmission interfaces.

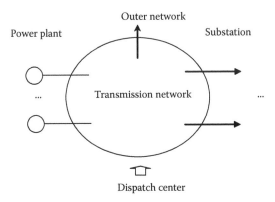

Figure 2.23 Power system operation.

- Day ahead: conduct load forecasting, generation scheduling, and transmission outage management operations.
- Real time: perform real-time economic dispatch.
- After event: perform data acquisition and settlement analysis.

In this chapter, we introduced fundamental concepts and mathematical models for economical dispatch, AGC, reserve, and generation scheduling. In practical applications, the mathematical models are more complex but the basic concepts exhibit few differences. For example, in practical power system operation, transient stability constraints should be considered. Other important operation concepts such as short-term load forecasting, state estimation, and maintenance scheduling are not covered in this book.

Appendix 2A: Determining Requirements of Ancillary Services

Because of the high cost of ancillary services, it is useful to determine the requirements (demands) of different kinds of reserves accurately. In this appendix, we introduce several basic methods for ancillary service determination.

Ten-Minute Reserve

A 10-minute reserve can be either a spinning or non-spinning type. Two methods exist for determining 10-minute reserve requirements: (1) the deterministic method ($N - 1$ rule) and (2) the probabilistic method. As an example of the deterministic method, the 10-minute reserve demand can be equal to 10% of the total load or the maximum capacity of the on-line generators.

Deterministic methods often yield conservative results. In essence, a system reserve is activated only after the occurrence of a random event. Therefore a probabilistic approach is usually more appropriate. Security and costs should be considered when determining the requirements for reserves. A heuristic method works as follows. Given a reserve requirement level, calculate the reserve constrained unit commitment problem, the corresponding reserve marginal cost, and the system reliability indices. Obviously, the greater the reserve, the higher the cost and the corresponding reliability of a system.

By repeating this process several times, we can determine a reserve requirement level that corresponds to a reasonable cost. This approach requires solving unit commitment problems many times and thus creates a rather high calculation burden. The Australia National Electricity Market and the PJM Electricity Market adopted this method. Many power dispatch centers argue that local transmission areas should be equipped with local reserves. In this chapter, we will not discuss this problem further.

Primary Frequency Reserve

A popular criterion for determining the primary frequency reserve requirement is that low frequency load shedding (LFLS) will not activate after a large disturbance of a system occurs. This criterion has gained application in North America and Australia and is in essence simply another form of the $N-1$ rule.

Now let us introduce a way to determine the primary frequency reserve requirement according to this criterion. Suppose the pickup frequency is f_{LS} for the first stage load shedding (LS). A short-term frequency greater than $f_{LS} + \Delta f$ should be ensured by providing fast reserves after a large disturbance occurs. Δf is the security margin according to operation experience. Suppose the load frequency characteristic of a system under study is

$$\frac{\partial Load}{\partial Frequency}$$

(this value can be measured). The primary reserve requirement can be determined through the following formula (assume the nominal frequency is 50 Hz):

$$reserve\ requirement = outage\ power - total\ load \times$$

$$\frac{\partial Load}{\partial Frequency} \times (50 - f_{LS} + \Delta f) \qquad (2.125)$$

Now suppose the pickup frequency for high frequency generator tripping (HFGT) is f_{GT}. Similarly, the system demand for fast down regulation reserve can be determined by

$$reserve\ requirement = surplus\ power - total\ load \times$$

$$\frac{\partial Load}{\partial Frequency} \times (f_{GT} - \Delta f - 50) \qquad (2.126)$$

Requirement for AGC Regulation Capacity

In some utilities, AGC regulation capacity requirements are based on operation experiences. For example, 1% of system total load is adopted as a requirement of AGC regulation capacity in Pennsylvania, New Jersey, and Maryland (www.pjm.com). Table 2.2 was developed for New England utilities (www.iso-ne.com). An empirical formula has been employed by the Czech Power Dispatch Company:

$$D_A = Round\left(\pm\sqrt{10 \times max\_load + 2250} - 150\right) + \varepsilon \qquad (2.127)$$

Table 2.2   New England AGC Regulation Capacity Requirements

Hour	Jan	Feb	Mar	Apr	May	Jun	Jul	Aug	Sep	Oct	Nov	Dec
1	590	590	510	300	300	450	560	560	450	450	590	590
2	430	430	350	240	240	350	400	400	250	250	430	430
3	410	410	300	200	200	270	330	330	250	250	410	410
4	390	390	300	200	200	270	310	310	250	250	390	390
5	420	420	300	200	200	270	310	310	250	250	420	420
6	600	600	450	300	300	420	500	500	450	450	600	600
7	700	700	590	360	360	550	580	580	500	500	700	700
8	780	780	740	460	460	600	680	680	550	500	780	780
9	740	740	690	460	460	600	660	660	500	500	740	740
10	690	690	630	400	400	530	640	640	500	500	690	690
11	550	550	500	340	340	450	510	510	420	420	550	550
12	530	530	470	340	340	450	510	510	420	420	530	530
13	520	520	440	340	340	450	510	510	420	420	520	520
14	490	490	430	340	340	450	530	530	420	420	490	490
15	480	480	390	340	340	450	540	540	420	420	480	480
16	550	550	460	340	340	500	650	650	420	420	550	550
17	580	580	460	340	340	500	650	650	420	420	580	580
18	710	710	510	360	360	500	680	680	450	550	710	710
19	760	760	670	360	360	500	630	630	500	500	760	760
20	780	780	700	360	360	500	600	600	500	500	780	780
21	760	760	680	450	450	500	630	630	550	550	760	760
22	760	760	650	450	450	550	660	660	550	550	760	760
23	800	800	670	450	450	600	730	730	550	550	800	800
24	680	680	640	400	400	550	650	650	550	550	680	680

where ε is an adjustment coefficient for system load variation. For example, during the morning periods of fast load increases, this coefficient may be larger. The details of this method are not illustrated in the literature.

The essence of AGC regulation capacity is a reserve capacity intended to satisfy the discrepancy between minute-level generation plan and load. When the power discrepancy is positive (negative), AGC up (down) regulation should be dispatched to increase (decrease) power output. No matter what direction the regulation follows, the AGC regulation capacity should at least exceed the power discrepancy.

Following the above principle, Texas ISO developed a systematic method to determine AGC requirements. The method is based on historical statistics of load variations. The maximum, minimum, and average values of load variations for specified 5-minute segments are saved and used to determine the AGC requirements for a period. Under this method, the AGC requirement should be greater than, for example, 99% of load variations. In other words, AGC reserve is required to cover load variations at a probability of 0.99.

The method employed in California ISO goes one step further by considering the violation times of AGC regulation criteria (CPS1 and CPS2) during a recent period such as a week and analyzing the actual usages of AGC regulation capacity. For example, if CPS2 has been violated many times or the practical usage of AGC is large, the AGC requirement should be increased to a certain extent and decreased if the reverse occurs.

We now describe the methodology in detail. In the first step, we acquire average, maximum, and minimum values of system total load, errors in load forecasting, and CPS2 data. Second, we prepare a table showing variations of system load for the past 2 years (Table 2.3). Assume that at probability 0.98 or higher the AGC regulation capacity can compensate load variations. Based on Table 2.3, the AGC up regulation capacity should be 300 MW. The AGC down regulation capacity can be calculated in a similar way. As an application of the method, in Texas, 300 MW is treated as the up regulation AGC capacity (Figure 2.24). In the third step, we revise the results from the second step based on the following rules:

Table 2.3 Positive Load Variations

Positive Load Deviation (MW)	Ratio of Positive Load Deviation to Total Sampling Times (%)
100	95
200	96
300	98
400	99

Note: A separate table is needed to include negative load variations.

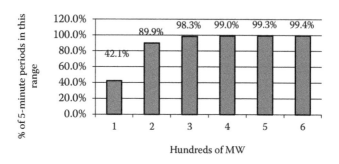

Figure 2.24 **Principle for determining up-regulation AGC capacity.**

■ If the CPS2 was violated once in the past week, we increase total AGC regulation capacity by 1%.

■ If the practical usage of AGC regulation capacity is (for example) N MW, the AGC requirement should be greater than $N \times 120\%$ MW.

We will use the Zhejiang power grid as an example. The load statistics table was built from historical data of the grid from 2000 to 2002. The AGC regulation capacity of the grid was forecast on December 5, 2002. The load curve for the day is shown in Figure 2.25. The forecast AGC up and down regulation capacities are shown in Figure 2.26.

As can be seen from the figure, the variation pattern of the procured AGC up and down regulation capacity is closely related to the load variation pattern. For example, from 8:30 to 11:30 a.m. and from 3:00 to 5:00 p.m. the requirements for up and down regulation are larger. During the other periods, the requirements are relatively smaller and in fact flatter. This general pattern is consistent with operational experiences.

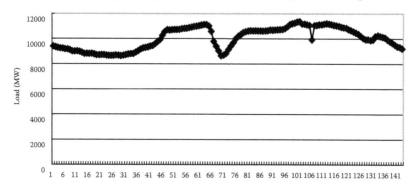

Figure 2.25 **System load curve of Zhejiang power grid on December 5, 2002.**

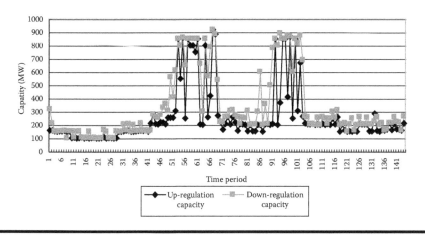

Figure 2.26 Forecast up- and down-regulation capacities of Zhejiang power grid on December 5, 2002.

Appendix 2B: Constraint Regularity Conditions for Non-Linear Programming

With the application of the Sard theorem, we show that constraint regularity conditions for non-linear programming are almost always satisfied (Spingarn and Rockafellar 1979). First, let us introduce several basic concepts.

If for i = 1, 2, ..., n, we have $a_i < b_i$, the set $(a_1, b_1) \times (a_2, b_2) \times ... \times (a_n, b_n)$ is an open rectangle. The volume of a rectangle can be denoted as $vol[(a_1, b_1) \times (a_2, b_2) \times ... \times (a_n, b_n)]$, which equals

$$\prod_{i=1}^{n} (b_i - a_i)$$

A set $C \subset R^n$ covered by open rectangles whose volumes are arbitrarily small is called a measure zero set. In other words, for arbitrarily small ε, there is an open rectangular series $R_1, R_2, ...$ in R^n that ensures

$$C \subset \bigcup_{i=1}^{\infty} R_i, \quad \sum_{i=1}^{\infty} vol(R_i) < \varepsilon$$

C is a measure zero set in R^n. As an example, a set of isolated points in R^n is a measure zero set, and R^k ($k < n$) is also a measure zero set in R^n.

Let $\mathbf{F}: B \to R^n$ be a continuously differentiable function where $B \subset R^q$. Consider the following n simultaneous equation with q variables: $\mathbf{F(x)} = \mathbf{e}$.

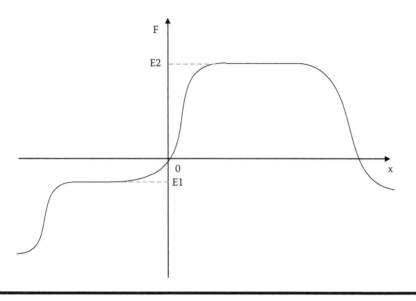

Figure 2.27 Function F.

Obviously, the above simultaneous equations generally have multiple solutions. Let the solution set be F_e^{-1}. We say that for a given right side term **e**, the function **F** is *regular* if the rank of matrix $\partial F/\partial x$ is equal to n at $x \in F_e^{-1}$. Figure 2.27 clearly shows that for all e except E_1 and E_2, the function **F** is *regular*. This is what the Sard theorem asserts.

Sard Theorem

Suppose $F: D \subset R^q \to R^n$ is a differentiable function where D is a closed set. **F** is *regular* except on a measure zero set. For proof of the above result, readers should consult a textbook on differential geometry. This theorem demonstrates that **F** is regular for almost all of **e**. Now, let us look at another optimization problem:

$$Min \quad c(\mathbf{x})$$

$$S.T. \quad \mathbf{g}(\mathbf{x}) = \mathbf{d}$$

Suppose $\mathbf{g}(\mathbf{x})$ is a continuously differentiable function. The Sard theorem tells us that, for almost all vector **d**, the above optimization problem satisfies constraint regularity conditions.

References

M.S. Bazaraa, H.D. Sherali, and C.M. Shetty. 1993. *Nonlinear Programming: Theory and Algorithms*, 2nd ed. New York: John Wiley & Sons.

G. Hamoud. 2000. Assessment of available transfer capability of transmission systems. *IEEE Transactions on Power Systems*, 15, 27–32.

E. Hirst and B. Kirby. 1996. Costs for electric power ancillary services. *Electricity Journal*, 9.

W. Li, P. Wang, and Z. Guo. 2006. Determination of optimal total transfer capability using a probabilistic approach. *IEEE Transactions on Power Systems*, 21, 862–868.

J.E. Spingarn and R.T. Rockafellar. 1979. The generic nature of optimal conditions in nonlinear programs, *Mathematics of Operations Research*, 4, 425–430.

Bibliography

J. Arrillaga and N.R. Watson. 2001. *Computer Modeling of Electrical Power Systems*, 2nd ed. New York: John Wiley & Sons.

K.W. Cheung, P. Shamsollahi, D. Sun et al. 2000. Energy and ancillary service dispatch for the interim ISO New England electricity market. *IEEE Transactions on Power Systems*, 15, 968–974.

A.I. Cohen and V.R. Sherkat. 1987. Optimization-based methods for operations scheduling. *Proceedings of IEEE*, 75.

D. Gan, X. Luo, D.V. Bourcier et al. 2003. Min–max transfer capability of a transmission interface. *International Journal of Electrical Power and Energy Systems*, 25.

N.S. Rau, 1999. Optimal dispatch of a system based on offers and bids: a mixed integer LP formulation. *IEEE Transactions on Power Systems*, 14, 274–279.

B. Stott, O. Alsac, and A.J. Monticelli. 1987. Security analysis and optimization. *Proceedings of IEEE*, 75.

A.J. Wood and B.F. Wollenberg. 1984. *Power Generation, Operation, and Control*, 2nd ed. New York: John Wiley & Sons.

X. Zhao, F. Wen, D. Gan et al. 2004. Determination of AGC capacity requirement and dispatch considering performance penalties. *Electric Power Systems Research*, 70, 93–98.

Chapter 3

Market Design: Spot Energy Market

As noted in Chapter 1, market mechanisms are expected to produce more social benefits. However, developing an efficient market is not as easy as it seems. In the United States, the airline and telecommunication industries achieved remarkable success after marketization around 1980 and 1995, respectively. We see continued decreases of air fares and communication charges while electricity has shown no signs of decreased prices since marketization in 1996. Certain special features of power systems are critical factors for electricity market design:

- High storage costs
- Transmission restriction and loss
- Lack of one-to-one correspondence between contracts and load flows
- Centralized coordination

Based on these factors, open bidding appears a natural step. The basic principle for a pool-based electricity market is to build one organization in charge of all the bidding and tendering affairs. The operation method of a pool-based electricity market is similar to that of a traditional power system. Thus, a pool-based electricity market is considered the best transitional market model.

In fact, pool-based electricity markets also allow bilateral trading to some extent. Many researchers, especially economists in the U.S., believe that a pool-based market is simply the best model of all. Our book also focuses on pool-based electricity markets.

Other economists argue that bilateral trading is the best market model because it allows both trading sides to negotiate and price freely to gain the benefits of free competition and makes the market transparent. However, the real-time operating characteristics of power systems may require balancing markets. The United Kingdom pioneered the bilateral trading electricity market in 2001.

Generally, we must solve some basic problems before developing an electricity market. What organizations are necessary? How should we operate the system? How will we settle accounts? Who is responsible for transmission and distribution fees? This chapter will answer the first three questions by introducing some popular market models or structures. The last question will be covered in the next two chapters.

3.1 Organization after Deregulation

The traditional electric power industry operates in a monopoly mode that combines generation, transmission, distribution, and sales within a single system. Electricity markets allow free competition on the generation side, while the transmission and distribution sides are still under government control according to the constant rate of return theory. With or without electricity markets, a power system must be operated by a single operation and coordination center, traditionally the system operator. Despite the development and expansion of electricity markets, operation and trade are still under the charge of one system operator or handled by different business organizations.

We now introduce some power industry entities and their major functions. A market operator (MO) or power exchange (PX) provides a trading place, matches supply and demand quotes, and sets energy prices. A system operator (SO) is an operational organization that functions independently of other interest groups. The transmission owner (TO) owns transmission equipment. A transmission system may be owned by several transmission owners (in the U.S.) or by one owner (in the U.K. and China).

The organization structures in various countries differ widely, as shown in Figure 3.1. The figure indicates that the market operator and system operator are integrated while the transmission owner is independent. This is the case in most U.S. electricity markets, for example, the PJM interconnection system and markets in New York State and New England. The electricity market in California adopted this structure after 2002. The Federal Energy Regulatory Commission (FERC) in the U.S. recently promoted centralized management of transmission owners, market operators, and system operator—a system China and the U.K. continue to use.

The supervision of electricity markets is the duty of independent system operators in the U.S. They are responsible for approving price restrictions, restricting bidding, auditing operation costs of generating units under price restrictions, and

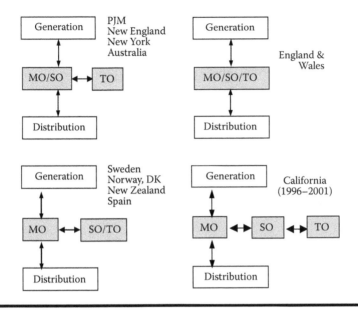

Figure 3.1   Organization of electricity markets.

writing market operation reports. The management groups of independent system operators are responsible to FERC, state legislatures, and market members.

Until now, retail electric prices in most electricity markets were subject to strict government restrictions. This type of restriction is known as "half-way deregulation." For example, consumers buy electricity from one utility. The utility's cost of buying power fluctuates but its selling price cannot be altered and this may pose a serious financial risk. Because of restricted selling prices, the Pacific Gas & Electric Company and Southern California Edison in California were almost on the verge of bankruptcy due to an abrupt escalation of bid prices in 2000.

The North American Electric Reliability Council (NERC) is responsible for formulation and supervision of calculation standards of reliability, automatic generation control (AGC), and other operational issues. In a mature electricity market, important data such as electric prices, load forecast results, market regulations, system planning principles, transmission fees, and maintenance scheduling should be released on the Internet and bidding data should be kept private.

3.2 Uniform Pricing

Traditionally, system operators process economic dispatch or security-constrained economic dispatch based on generating unit cost curves. In the electricity market environment, generating unit bidding price curves are used for the same purpose.

Electric utilities generally request incremental bidding price curves from units. The bidding price curve may be a zero curve if a bilateral trading contract exists between a unit and the load side, so we can say that pool-based electricity markets allow bilateral trading to some extent.

3.2.1 Model for Uniform Pricing

Assume that the bidding price curve of a generating unit j is a straight line with height p_j. Vector $\mathbf{p}$ denotes all bidding prices and total generation cost is represented by $\mathbf{p}^T\mathbf{P}_G$. Without the constraints of grid flows, the model for economic dispatch could be simplified as

$$\underset{\mathbf{P}_G}{Min} \quad \mathbf{p}^T\mathbf{P}_G \tag{3.1}$$

$$S.T. \quad \mathbf{e}^T\mathbf{P}_G = P_D \tag{3.2}$$

$$\underline{\mathbf{P}}_G \le \mathbf{P}_G \le \overline{\mathbf{P}}_G \tag{3.3}$$

Without considering minimum generation output, this problem may be solved using a simple ranking method. Practically, we first turn on the generating unit at minimum bidding price, then at the second minimum bidding price, and continue until the load is satisfied. This is illustrated in Figure 3.2. The units in shadow win the bid and the bidding price of the last unit becomes the final clearing price. Ranking is applied widely in electricity markets and readers should be familiar with the method. When generating units' minimum outputs are taken into consideration, dispatch plans may be carried out by a similar method.

Now we consider the settlement problem. The Lagrange function of the above optimization problem is

$$\Gamma = \mathbf{p}^T\mathbf{P}_G + \lambda(\mathbf{e}^T\mathbf{P}_G - P_D) + \sum_i \hat{\tau}_i(P_{Gi} - \overline{P}_{Gi}) - \sum_i \check{\tau}_i(P_{Gi} - \underline{P}_{Gi}) \tag{3.4}$$

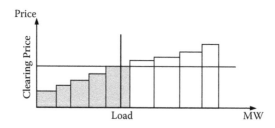

Figure 3.2   Ranking method for clearing price.

One of the optimization conditions may be expressed by

$$\frac{\partial \Gamma}{\partial P_{Gi}} = p_i + \lambda + \hat{\tau}_i - \check{\tau}_i = 0, \ (i = 1, 2, \ldots, N_G) \tag{3.5}$$

We can see from Figure 3.2 that only one generating unit usually outputs between the minimum and maximum output limits. We designate it the marginal generating unit and assign index m to it. According to Karush–Kuhn–Tucker conditions, the Lagrange multiplier for the marginal generating unit is $\check{\tau}_m = \hat{\tau}_m = 0$; therefore:

$$p_m + \lambda = 0 \tag{3.6}$$

For generating units reaching their maximum output limits, $\check{\tau} = 0$ and thus

$$p_i + \lambda + \hat{\tau}_i = 0 \text{ or } p_i + \hat{\tau}_i = -\lambda \tag{3.7}$$

because $\hat{\tau}_i > 0$, $p_i < -\lambda$, which means that the market clearing price is higher than the bidding price. From the view of a generating unit, $-\lambda$ is a reasonable settlement price because it will not cause discontent from marginal generating units, full output units, or zero output units. It is designated the marginal price or clearing price ρ, which is

$$\rho = -\lambda \tag{3.8}$$

Let us again study clearing price ρ, this time from the load side. We first introduce the envelope theorem for optimization. This theorem provides an easy method for analyzing optimization problems with parameters. Consider the following optimization:

$$f(a) = \underset{\mathbf{x}}{Min} \ c(\mathbf{x}, a) \tag{3.9}$$

$$S.T. \quad h_i(\mathbf{x}, a) = 0, \ (i = 1, 2, \ldots, l) \tag{3.10}$$

with a Lagrange function:

$$\Gamma = c(\mathbf{x}, a) - \sum_{i=1}^{l} \lambda_i h_i(\mathbf{x}, a) \tag{3.11}$$

If $f(a)$ is differentiable, given a_0, we obtain an optimization solution $\mathbf{x}_0$. We claim that

$$\left. \frac{df}{da} \right|_{a=a_0} = \left. \frac{\partial \Gamma}{\partial a} \right|_{a=a_0} = \frac{\partial c(\mathbf{x}_0, a_0)}{\partial a} - \sum_{i=1}^{k} \lambda_i \frac{\partial h_i(\mathbf{x}_0, a_0)}{\partial a} \tag{3.12}$$

To prove this theorem, we first notice the constraint conditions expressed as

$$h_i\left(\mathbf{x}(a), a\right) = 0, (i = 1, 2, ..., l) \tag{3.13}$$

We differentiate both sides with a:

$$\frac{\partial h_i(\mathbf{x}, a)}{\partial \mathbf{x}} \cdot \frac{d\mathbf{x}(a)}{da} = -\frac{\partial h_i(\mathbf{x}, a)}{\partial a}, (i = 1, 2, ..., l) \tag{3.14}$$

We then multiply both sides with λ_i and summate. At $a = a_0$ we have

$$\sum_{i=1}^{l} \lambda_i \frac{\partial h_i(\mathbf{x}_0, a_0)}{\partial \mathbf{x}} \cdot \frac{d\mathbf{x}(a_0)}{da} = \sum_{i=1}^{l} -\lambda_i \frac{\partial h_i(\mathbf{x}_0, a_0)}{\partial a} \tag{3.15}$$

and build a Lagrange function:

$$\Gamma = c\left(\mathbf{x}, a\right) + \sum_{i=1}^{l} \lambda_i h_i\left(\mathbf{x}, a\right) \tag{3.16}$$

When $a = a_0$, the related optimization solution x_0 satisfies the extreme conditions and thus:

$$\frac{\partial c\left(\mathbf{x}_0, a_0\right)}{\partial \mathbf{x}} = \sum_{i=1}^{l} -\lambda_i \frac{\partial h_i\left(\mathbf{x}_0, a_0\right)}{\partial \mathbf{x}} \tag{3.17}$$

From the definition we have

$$f(a) = c(\mathbf{x}(a), a) \tag{3.18}$$

Thus at $a = a_0$:

$$\left.\frac{df}{da}\right|_{a = a_0} = \frac{\partial c\left(\mathbf{x}_0, a_0\right)}{\partial \mathbf{x}} \cdot \frac{d\mathbf{x}(a_0)}{da} + \frac{\partial c\left(\mathbf{x}_0, a_0\right)}{\partial a} \tag{3.19}$$

Taking Equations 3.17 to 3.19, we have

$$\left.\frac{df}{da}\right|_{a = a_0} = \sum_{i=1}^{l} \lambda_i \frac{\partial h_i\left(\mathbf{x}_0, a_0\right)}{\partial \mathbf{x}} \cdot \frac{d\mathbf{x}(a_0)}{da} + \frac{\partial c\left(\mathbf{x}_0, a_0\right)}{\partial a} \tag{3.20}$$

Taking Equations 3.16 to 3.20, we have

$$\left.\frac{df}{da}\right|_{a\,=\,a_0} = \sum_{i=1}^{l} -\lambda_i\,\frac{\partial h_i(\mathbf{x}_0, a_0)}{\partial a} + \frac{\partial c(\mathbf{x}_0, a_0)}{\partial a} = \left.\frac{d\Gamma}{da}\right|_{a\,=\,a_0} \tag{3.21}$$

The theorem is thus proven. Analyzing the influence of load on marginal price based on the envelope theorem, we have

$$\left.\frac{dc}{dP_D}\right|_{P_D} = \left.\frac{\partial \Gamma}{\partial P_D}\right|_{P_D} = -\lambda = \rho \quad \left(P_D = \sum P_{Gi}\right) \tag{3.22}$$

Equation 3.22 shows that the marginal price is relevant to load demand P_D, where

$$\left.\frac{dc}{dP_D}\right|_{P_D}$$

indicates the incremental cost as load P_D increases by one unit. ρ is also called the shadow price. We can see this relationship easily in the clearing price (Figure 3.2). This method is sometimes employed to verify marginal price equation, and readers should be familiar with it. Details about the envelope theorem may be found in Luenberger (1995). Interested readers may also refer to Schweppe et al. (1988) for information about the nature of marginal price from the view of utility maximization.

Example 3.1 — Assume a system has three generating units with separate bidding prices of $10, $20, and $30 per MWh. The unit capacities are 100, 200, and 300 MW. The minimum unit output is zero and the system load is 250 MW. Solve the clearing price.

Solution — We build the dispatch model as follows:

$$\operatorname*{Min}_{\mathbf{P}_G} \quad 10P_{G1} + 20P_{G2} + 30P_{G3}$$

$$S.T. \quad P_{G1} + P_{G2} + P_{G3} = 250$$

$$0 \le P_{G1} \le 100$$

$$0 \le P_{G2} \le 200$$

$$0 \le P_{G3} \le 300$$

The Lagrange function of this problem is

$$\Gamma = 10P_{G1} + 20P_{G2} + 30P_{G3}$$

$$+ \lambda(P_{G1} + P_{G2} + P_{G3} - 250)$$

$$+ \hat{\tau}_1(P_{G1} - 100) + \hat{\tau}_2(P_{G2} - 200) + \hat{\tau}_3(P_{G3} - 300)$$

$$- \breve{\tau}_1 P_{G1} - \breve{\tau}_2 P_{G2} - \breve{\tau}_3 P_{G3}$$

The optimization condition is

$$10 + \lambda + \hat{\tau}_1 - \breve{\tau}_1 = 0, \ \ 20 + \lambda + \hat{\tau}_2 - \breve{\tau}_2 = 0, \ \ 30 + \lambda + \hat{\tau}_3 - \breve{\tau}_3 = 0$$

$$P_{G1} + P_{G2} + P_{G3} = 250$$

$$\hat{\tau}_1(P_{G1} - 100) = 0, \ \ \breve{\tau}_1 P_{G1} = 0$$

$$\hat{\tau}_2(P_{G2} - 200) = 0, \ \ \breve{\tau}_2 P_{G2} = 0$$

$$\hat{\tau}_3(P_{G3} - 300) = 0, \ \ \breve{\tau}_3 P_{G3} = 0$$

The optimization results are $P_{G1} = 100$, $P_{G2} = 150$, $P_{G3} = 0$, and $\rho = -\lambda = 20$. The Lagrange multipliers $\hat{\tau}$ and $\breve{\tau}$ may be calculated also and their significance is clear. Example 3.1 assumes that the minimum output for each generating unit is zero, but actually the thermal generator minimum outputs are greater than zero. In this case, some good properties of uniform pricing may be damaged. For example, assume the minimum output of unit 3 is 20 MW; then

$$\underset{P_G}{Min} \quad 10P_{G1} + 20P_{G2} + 30P_{G3}$$

$$S.T. \quad P_{G1} + P_{G2} + P_{G3} = 250$$

$$0 \le P_{G1} \le 100$$

$$0 \le P_{G2} \le 200$$

$$20 \le P_{G3} \le 300$$

Accordingly, the optimization solutions changed to $P_{G1} = 100$, $P_{G2} = 130$, $P_{G3} = 20$, and $\rho = -\lambda = 20$. Since the $30 bidding price of unit 3 is higher than the marginal price of $20, the extra compensation payment is called the uplift fee.

The main drawback of uniform pricing is its failure to embody the value of positioning for generating units and loads. Using the two-generator system in Figure 3.3 as an example, unit G1 with a lower bidding price is "constrained down"

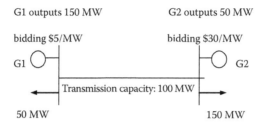

Figure 3.3 Two-generator system.

due to a transmission capacity constraint. We have two market pricing methods. One method uses \$5/MW as a clearing price. Meanwhile, unit G2 still has a \$30/MW bidding rate that includes a congestion uplift fee of $\$(30 - 5) \times 50 = \$1250$. The congestion uplift fee is proportionally allocated to all loads and is obviously unfair to the load on G1's side. This demonstrates the most significant problem of uniform pricing. The other method utilizes \$30 as its clearing price. In this case, the congestion uplift fee no longer exists. However, unit G1 may think the price unfair because it is constrained down despite lower bidding.

3.2.2 Bilateral Trading in Pool-Based Electricity Markets

Assume that a bilateral trading contract is reached between generating unit A and a load. The contract MW requirement and price are specified. In a pool-based market, the unit should bid at a price low enough for this amount of electricity to ensure that at least the contract MW wins the bid and the contract can be realized. In this case, the load pay is calculated as contract price × contract MW. The generating unit receives contract price × contract MW + CP × (online WM − contract MW).

In China, free competition in the electricity market is unacceptable for many reasons. For example, old power plants may have passed the loan period and thus can operate at low marginal cost, while new power plants may be still under pressure from loans and thus operate at high marginal cost. Some new power plants have complicated financial arrangements that include special generation contracts. Therefore, it is unrealistic to force all old and new plants to bid together in a pool-based electricity market.

Furthermore, power systems are so complicated that regional resolutions should be attempted before free competition is introduced. This is why the contract for difference (CfD) concept was introduced as part of electric industry deregulation in China. The basic motivation is to allow competition in a segment of total electricity and allocate other electricity according to a planned economy. Thus the formula for the income of a power plant is generation income = market selling income + contract for difference income, or

$$Revenue = P_G \cdot \rho + P_C \cdot (\rho_c - \rho) = P_C \cdot \rho_c + \rho \cdot (P_G - P_C) \quad (3.23)$$

The contract for difference price ρ_C and electricity P_C are determined in the contract between the system operator and the plant and the arrangement is supervised by local government. P_G is the cleared output in the market and ρ is the market clearing price. We may conclude from the above equation that (1) when online electricity equals contract electricity ($P_G = P_C$), the unit receives contract income $P_C\rho_C$; (2) when online electricity exceeds contract electricity ($P_G > P_C$), the income of the unit exceeds what it receives from contract income; and (3) when online output is smaller than contract output ($P_G < P_C$), the income of the unit is less than what it receives from contract income.

Therefore, one generating unit needs only to bid at a price low enough to ensure an income equal to or larger than contract income. This is how CfD protects generation companies and electricity sellers, and utility companies and electricity buyers avoid market risk. This kind of protection may be helpful during the gradual introduction of electricity markets.

3.3 Nodal Pricing

3.3.1 Model for Nodal Pricing

The main problem of uniform electricity pricing is its ineffectiveness in dealing with pricing under grid congestion conditions. As an alternative, nodal pricing will be discussed in this section. Nodal pricing is widely applied worldwide because it can deal with grid congestion naturally via optimal load flow techniques and reasonably reflects the marginal cost at load node.

The nodal price at a certain node is defined as the incremental cost for the system to provide the node with one additional megawatt of electricity. Figure 3.4 shows that when the load at node 1 increases by 1 MW, unit G1 will accordingly increase output by 1 MW and the electricity supply cost of the system will increase by $5. This makes the nodal price $5/MW. Similarly, the nodal price at node 2 is

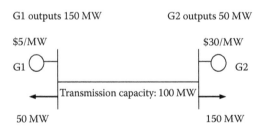

Figure 3.4 Two-generator system and marginal prices.

$30/MW. As to settlement, G1 and load 1 use the nodal price at node 1, while G2 and load 2 use the nodal price at node 2.

We now introduce the model for nodal pricing. The system operator first conducts the following economic dispatch:

$$\underset{\mathbf{P}_G}{Min} \quad \mathbf{p}^T \mathbf{P}_G \tag{3.24}$$

$$S.T. \quad \mathbf{e}^T (\mathbf{P}_G - \mathbf{P}_D) = 0 \tag{3.25}$$

$$\mathbf{T}(\mathbf{P}_G - \mathbf{P}_D) \leq \overline{\mathbf{F}} \tag{3.26}$$

$$\underline{\mathbf{P}}_G \leq \mathbf{P}_G \leq \overline{\mathbf{P}}_G \tag{3.27}$$

The Lagrange function for this optimization problem is

$$\Gamma = \mathbf{p}^T \cdot \mathbf{P}_G + \lambda \mathbf{e}^T (\mathbf{P}_G - \mathbf{P}_D) + \mathbf{\mu}^T [\mathbf{T}(\mathbf{P}_G - \mathbf{P}_D) - \overline{\mathbf{F}}] + \\ \widehat{\mathbf{\tau}}^T (\mathbf{P}_G - \overline{\mathbf{P}}_G) + \widecheck{\mathbf{\tau}}^T (\underline{\mathbf{P}}_G - \mathbf{P}_G) \tag{3.28}$$

The first-order optimization condition is

$$\frac{\partial \Gamma}{\partial P_{Gi}} = p_i + \lambda + \sum_k \mu_k T_{ki} + \widehat{\tau}_i - \widecheck{\tau}_i = 0 \quad \left(i = 1,\ 2,\ NG \right) \tag{3.29}$$

The nodal price for node i is

$$\rho_i = -\lambda - \sum_k \mu_k T_{ki} \tag{3.30}$$

The above equation may be expressed in vector form as

$$\mathbf{\rho} = -\lambda \mathbf{e} - \mathbf{T}^T \mathbf{\mu} \tag{3.31}$$

Based on Equation 3.31, we can analyze electric prices for different generating units. For marginal generating unit i, $\widehat{\tau}_i = \widecheck{\tau}_i = 0$, so that $p_i = \rho_i$, meaning its bidding price equals its nodal price. For generating unit i operating at its maximum output limit, $\widecheck{\tau}_i = 0$, meaning that $p_i - \rho_i + \widehat{\tau}_i = 0$ or $p_i + \widehat{\tau}_i = \rho_i$. Since $\widehat{\tau}_i > 0$, we

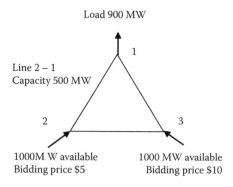

Figure 3.5 Nodal price for a three-node system.

have $p_i < \rho_i$. For the unit operating at its minimum output limit, $\hat{\tau}_i = 0$, meaning $p_i - \rho_i - \breve{\tau}_i = 0$, or $p_i - \breve{\tau}_i = \rho_i$. Since $\breve{\tau}_i > 0$, we have $p_i > \rho_i$. From this view, nodal marginal price is a reasonable settlement price because, in this case, all marginal units, full output units, and zero output units are satisfied.

In a nodal pricing system, for a generating unit operating at its minimum output limit, we assume its output to be $\underline{\mathbf{P}}_G$. Generally, $\underline{\mathbf{P}}_G \neq 0$—the same as in the uniform pricing system. In this case, the units should receive an uplift payment at the rate of their bidding prices. Because their bidding prices are higher than the nodal price, the uplift fees should be allocated to all consumers.

We will analyze the marginal price again from the load side. The objective function of the optimization problem involves the total electricity purchase fee $c = \mathbf{p}^T \mathbf{P}_G$ that varies with load $\mathbf{P}_D$. According to the envelope theorem:

$$\frac{\partial c}{\partial \mathbf{P}_D} = \frac{\partial \Gamma}{\partial \mathbf{P}_D} = -\lambda \mathbf{e} - \mathbf{T}^T \boldsymbol{\mu} = \boldsymbol{\rho} \tag{3.32}$$

The equation shows that when the load of a certain system increases by 1 MW, its electric purchase fee increases by its nodal price. This is a positive feature because it conforms to the concept that the "one who uses high priced electricity pays a high price." This method may be used to examine the nodal price of the two-node system illustrated in Figure 3.5.

Example 3.2 — Analyze the mathematical model and price for the three-node system illustrated in Figure 3.5. Note that the final nodal price of node 1 is higher than 10, which is the highest bidding price among the generating units.
Solution — The mathematical model of this problem is

$$\underset{\mathbf{P}_G}{Min} \quad 5P_{G2} + 10P_{G3}$$

$$S.T. \quad P_{G2} + P_{G3} - 0.9 = 0$$

$$\begin{bmatrix} -0.3333 & 0.3333 & 0 \\ -0.6667 & -0.3333 & 0 \\ 0.3333 & 0.6667 & 0 \end{bmatrix} \begin{bmatrix} -0.9 \\ P_{G2} \\ P_{G3} \end{bmatrix} \le \begin{bmatrix} \overline{F}_{21} \\ \overline{F}_{31} \\ \overline{F}_{23} \end{bmatrix}$$

$$0 \le P_{G2} \le 1$$

$$0 \le P_{G3} \le 1$$

Assume that the transmission capacities of lines 3-1 and 2-3 are all high enough so that the inequality constraints relating to these two lines may be ignored. The results are: P2 = 0.6 and P3 = 0.3; Price 1 = 15.0, Price 2 = 5.0, and Price 3 = 10.0. The results show that the nodal price of node 1 is $15/MW. In the above model, the objective function is to minimize production cost.

When the two generating units in the above problem both output 450 MW, the settlement price for the entire grid will be $10, which means the consumer's price is less than the nodal price. However, nodal pricing is still preferable because it minimizes social production cost. Although the idea that both generators output 450 MW may reduce consumer prices, it is unreasonable because it increases the production cost over the entire grid.

As discussed above, one advantage of nodal pricing is its adherence to the common sense principle that the user of high priced electricity pays a high price. However, this advantage will be somewhat dampened for generating units with small ramp rates or long start-up times. The three-generator system in Figure 3.6

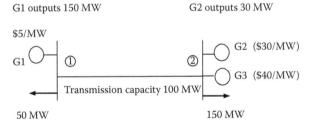

G1 outputs 150 MW   G2 outputs 30 MW

Minimum output of G3 is 20 MW it thus outputs 20 MW
Nodal prices for nodes 1, 2 are $5, $30/MW

Figure 3.6   Uplift fee in nodal pricing ($200).

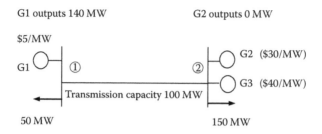

G1 outputs 140 MW   G2 outputs 0 MW

$5/MW

G1 ①   ② G2 ($30/MW)

G3 ($40/MW)

Transmission capacity 100 MW

50 MW   150 MW

Nodal price nodes 1, 2 are all $5/MW
Minimum output of G3 is 60 MW thus it outputs 60 MW

Figure 3.7   Uplift fee in nodal pricing ($175).

is an example. Unit G3 has a non-zero minimum output of 20 MW. According to economic dispatch, its output should be 20 MW. Meanwhile, unit G2 outputs 30 MW and is the marginal unit. Thus its bidding price is used to set the nodal price for node 2, which is $30/MW. As the bidding price of unit G3 is higher than the nodal price, it receives an uplift fee of $(40 − 30) × 20 = $200. This uplift fee is usually allocated to all loads and thus minimizes the advantage of nodal pricing.

The situation could be worse. Assume that the minimum output of G3 is 60 MW and it should be 60 MW according to economic dispatch as in Figure 3.7. G1 is the only marginal generating unit and the system price is the bidding price of G1, which is $5/MW. As the bidding price of G3 is much higher, the uplift fee would be $(40 − 5) × 5 = $175 allocated to all loads. In this case, the minimum output of G3 eliminates the nodal price difference between nodes 1 and 2 and the advantage of nodal pricing no longer exists.

3.3.2 Selection of a Reference Node

If the transmission loss is not considered, will a different reference node bring about a different nodal price? The answer is no. Assume that the Power Transfer Distribution Factor (PTDF) matrix is T' under a certain reference node. The dispatch model is

$$\underset{P_G'}{Min} \quad f' = \mathbf{p}^T \mathbf{P}_G' \tag{3.33}$$

$$S.T. \quad \mathbf{e}^T (\mathbf{P}_G' - \mathbf{P}_D) = 0 \tag{3.34}$$

$$\mathbf{T}'(\mathbf{P}_G' - \mathbf{P}_D) \le \overline{\mathbf{F}} \tag{3.35}$$

$$\underline{\mathbf{P}}_G \le \mathbf{P}_G' \le \overline{\mathbf{P}}_G \tag{3.36}$$

When the reference node is changed, the new PTDF matrix is $\mathbf{T}''$ and the new dispatch model is

$$\underset{\mathbf{P}_G''}{Min} \quad f'' = \mathbf{p}^T \mathbf{P}_G'' \tag{3.37}$$

$$S.T. \quad \mathbf{e}^T (\mathbf{P}_G'' - \mathbf{P}_D) = 0 \tag{3.38}$$

$$\mathbf{T}''(\mathbf{P}_G'' - \mathbf{P}_D) \leq \overline{\mathbf{F}} \tag{3.39}$$

$$\underline{\mathbf{P}}_G \leq \mathbf{P}_G'' \leq \overline{\mathbf{P}}_G \tag{3.40}$$

Obviously, the selection of reference node will not influence load flow, so we have

$$\mathbf{T}''(\mathbf{P}_G'' - \mathbf{P}_D) = \mathbf{T}'(\mathbf{P}_G'' - \mathbf{P}_D) \tag{3.41}$$

By substituting Equation 3.41 to 3.39, we obtain:

$$\underset{\mathbf{P}_G''}{Min} \quad \mathbf{p}^T \mathbf{P}_G'' \tag{3.42}$$

$$S.T. \quad \mathbf{e}^T (\mathbf{P}_G'' - \mathbf{P}_D) = 0 \tag{3.43}$$

$$\mathbf{T}'(\mathbf{P}_G'' - \mathbf{P}_D) \leq \overline{\mathbf{F}} \tag{3.44}$$

$$\underline{\mathbf{P}}_G \leq \mathbf{P}_G'' \leq \overline{\mathbf{P}}_G \tag{3.45}$$

This model is exactly the same as the original except for the changed symbol. We may solve the two mathematical models ($\mathbf{P}_G'' = \mathbf{P}_G'$ and $f' = f''$) and obtain the same optimization solution. We now assign nodal prices under the two situations to variables $\boldsymbol{\rho}'$ and $\boldsymbol{\rho}''$. We know from the envelope theorem that

$$\boldsymbol{\rho}' = \frac{\partial f'}{\partial \mathbf{P}_D} = \frac{\partial f''}{\partial \mathbf{P}_D} = \boldsymbol{\rho}'' \tag{3.46}$$

We have thus proven that different reference nodes will yield the same nodal price. Readers may also wonder whether nodal price components $\lambda \mathbf{e}$ and $\mathbf{T}^T \boldsymbol{\mu}$ will change when we change the reference node. This time, the answer is yes because

$$\boldsymbol{\rho}' = \lambda' \mathbf{e} + \mathbf{T}'^T \boldsymbol{\mu}' \tag{3.47}$$

$$\boldsymbol{\rho}'' = \lambda'' \mathbf{e} + \mathbf{T}''^T \boldsymbol{\mu}'' \tag{3.48}$$

For reference node r, the nodal price

$$\rho_r = \lambda + \sum_{k=1}^{NL} T_{kr}\mu_k$$

As $T_{kr} = 0(k = 1, 2, ..., NL)$, λ' and λ'' will equal the prices of their reference nodes. Therefore, the difference between λ' and λ'' equals the price difference between reference nodes. Since $\rho' = \rho''$, the difference between congestion components also equals the price difference between reference nodes.

3.3.3 Sparse Form of Nodal Pricing

The above nodal pricing model employs a dense form of load flow model. We can also build a sparse form of economic dispatch model:

$$\underset{P_G,\theta}{Min} \quad \sum_{i=1}^{NG} c_i(P_{Gi}) \tag{3.49}$$

$$S.T. \quad \mathbf{P}_G - \mathbf{P}_D = \mathbf{B}\theta \tag{3.50}$$

$$\mathbf{X}\theta \le \overline{\mathbf{F}} \tag{3.51}$$

$$\underline{\mathbf{P}}_G \le \mathbf{P}_G \le \overline{\mathbf{P}}_G \tag{3.52}$$

We build the Lagrange function:

$$\Gamma = \mathbf{p}^T \cdot \mathbf{P}_G + \gamma^T(\mathbf{P}_G - \mathbf{P}_D - \mathbf{B}\theta) + \mu^T[\mathbf{X}\theta - \overline{\mathbf{F}}] + \hat{\tau}^T(\mathbf{P}_G - \overline{\mathbf{P}}_G) + \check{\tau}^T(\underline{\mathbf{P}}_G - \mathbf{P}_G) \tag{3.53}$$

The optimization condition is

$$\frac{\partial\Gamma}{\partial P_{Gi}} = p_i + \gamma_i + \hat{\tau}_i - \check{\tau}_i = 0 \quad (i = 1, 2, ..., N_G) \tag{3.54}$$

The nodal price for node i is

$$\rho_i = -\gamma_i \tag{3.55}$$

The variable $-\gamma_i$, $(i = 1, 2, N_G)$ defined in the above equation is sometimes called the nodal incremental cost. The nodal price derived from a sparse form of grid function is the same as that derived from a dense form. This may be proven using the envelope theorem.

3.4 Multiple Block Bidding

In the material above, we implicitly assume one generator bids only one block for simplicity. However, participants in real-world electricity markets can offer multi-block bids. Usually the block prices are required to be monotonically increasing. Figure 3.8 depicts a generator offer with multi-blocks. The width of each block indicates the bid capacity (of the block), while the height of each block indicates the bid price (of the block).

Let us now consider the dispatch model of multi-block bids. Let ϕ_i^j, Φ_i^j, and $\bar{\Phi}_i^j$ denote the bid price, cleared capacity, and bid capacity of the jth bid block of the ith generator, respectively. For simplicity, assume all generators bid the same number, i.e., $N\Phi$, of blocks. Since the bid curve is increasing, we have for any generator i ($i = 1, 2, ..., NG$):

$$\phi_i^1 \leq \phi_i^2 \leq ... \leq \phi_i^{N\Phi} \tag{3.56}$$

and at optimal solution,

$$\Phi_i^1 + \Phi_i^2 + ... + \Phi_i^{N\Phi} = P_{Gi} \tag{3.57}$$

Notice that the production cost of the ith generator is $\phi_i^T \Phi_i$. We then have the model of multi-block bidding:

$$\underset{P_G, \Phi}{Min} \quad \sum_{i=1}^{NG} \phi_i^T \Phi_i \tag{3.58}$$

$$S.T. \quad \Phi_i^1 + \Phi_i^2 + ... + \Phi_i^{N\Phi} = P_{Gi}, (i = 1, 2, ..., NG) \tag{3.59}$$

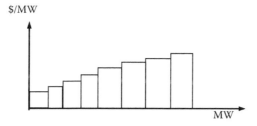

Figure 3.8   Generator offer with multi-blocks.

$$\mathbf{e}^T (\mathbf{P}_G - \mathbf{P}_D) = 0 \tag{3.60}$$

$$\mathbf{T}(\mathbf{P}_G - \mathbf{P}_D) \le \overline{\mathbf{F}} \tag{3.61}$$

$$\underline{\mathbf{P}}_G \le \mathbf{P}_G \le \overline{\mathbf{P}}_G \tag{3.62}$$

3.5 Demand Side Bidding

In the above uniform pricing model and nodal pricing model, demand side bidding is not taken into consideration. Actually, certain loads are dispatchable in a power system, for example, the load for a pumped storage unit during its pumping period. For a day-ahead market, other loads can also be dispatchable to a degree. A mature market should be designed to give loads the opportunity to bid. We will discuss such bidding briefly in this section.

Demand side bidding and generation side bidding are symmetrical in principle. Based on this conception, loads offer bidding price curves of their highest affordable prices to system operators. Generally, these curves are stepwise and decreasing. As shown in Figure 3.9, the two usual kinds of bidding price curves are capped and uncapped. In the case of capped curves, all loads are assigned a maximum affordable price. For the uncapped curves, some loads are not assigned a maximum affordable price, which means they are willing to pay at any rate.

During a period without congestion in an electricity market, the intersection of the load bidding price curve and the generating unit bidding price curve is the clearing price. Two clearing methods are illustrated in Figure 3.10. Assume that one node offers one bidding price curve and assign all loads a maximum affordable price. The real-time dispatch model is changed as demand side bidding is taken into consideration.

$$\underset{\mathbf{P}_G, \mathbf{P}_D}{Max} \quad \mathbf{p}_D^T \mathbf{P}_D - \mathbf{p}^T \mathbf{P}_G \tag{3.63}$$

$$S.T. \quad \mathbf{e}^T (\mathbf{P}_G - \mathbf{P}_D) = 0 \tag{3.64}$$

$$\mathbf{T}(\mathbf{P}_G - \mathbf{P}_D) \le \overline{\mathbf{F}} \tag{3.65}$$

$$\underline{\mathbf{P}}_G \le \mathbf{P}_G \le \overline{\mathbf{P}}_G \tag{3.66}$$

$$\underline{\mathbf{P}}_D \le \mathbf{P}_D \le \overline{\mathbf{P}}_D \tag{3.67}$$

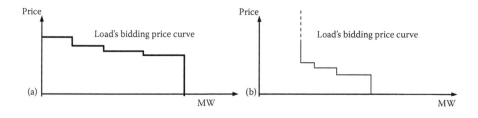

Figure 3.9 Load bidding price curves. (a) Capped load bidding price curve. (b) Uncapped load bidding price curve.

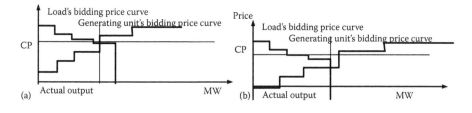

Figure 3.10 Two clearing methods. (a) According to load. (b) According to generating unit.

Applying the nodal pricing principles to the above model to analyze the electric price, we easily discover that the objective function is changed to maximize social welfare, instead of minimizing production cost without demand side bidding as in the previous model. The dispatch model for determining an uncapped load's bidding price curves is more complex, but the basic pricing principle remains the same.

3.6 Day-Ahead Market

3.6.1 Basic Principles

In day-ahead markets, generation scheduling and electric price calculation are similar to traditional dispatch situations. The only difference is the use of a bidding price curve instead of a cost curve. Some markets allow generating units to bid for electricity and also to bid for start-up and no-load operation fees. The process is called three-part bidding.

We first introduce generation scheduling and the pricing model of day-ahead markets. Generally in a pool-based electricity market, generating units can bid for start-up and shut-down fees. We use S_i^t to denote the start-up/shut-down fee of unit i ($i = 1, 2, ..., NG$) in dispatch time period t ($t = 1, 2, ..., 24$). c_i is the bidding price for electricity. The generation scheduling model of a day-ahead market is very similar to that introduced in Chapter 2:

$$\underset{\mathbf{P}_G,\mathbf{U}}{Min} \quad \sum_{i=1}^{NG}\sum_{t=1}^{24}\left[U_i^t c_i P_{Gi}^t + S_i^t U_i^t (1-U_i^{t-1})\right] \qquad (3.68)$$

$$S.T. \quad \mathbf{e}^T(\mathbf{P}_G^t - \mathbf{P}_D^t) = 0,\ (t=1,2,...,24) \qquad (3.69)$$

$$\mathbf{T}(\mathbf{P}_G^t - \mathbf{P}_D^t) \le \overline{\mathbf{F}},\ (t=1,2,...,24) \qquad (3.70)$$

$$U_i^t \underline{P}_{Gi}^t \le P_{Gi}^t \le U_i^t \overline{P}_{Gi}^t,\ (t=1,2,...,24)\ (i=1,2,...,NG) \qquad (3.71)$$

$$U_i^t \in \{0,1\},\ (t=1,2,...,24)\ (i=1,2,...,NG) \qquad (3.72)$$

Assume that the integer solutions of the above problem have been worked out. We thus know the on and off status of all units for all time periods. We then further divide the problem into different hours and calculate them independently. We use I to denote the set of all running units at moment t. The economic dispatch problem is thus changed to

$$\underset{\mathbf{P}_G,\mathbf{U}}{Min} \quad \sum_{i \in I}\left[c_i P_{Gi}^t + constant\right] \qquad (3.73)$$

$$S.T. \quad \mathbf{e}^T(\mathbf{P}_G^t - \mathbf{P}_D^t) = 0,\ (t=1,2,...,24) \qquad (3.74)$$

$$\mathbf{T}(\mathbf{P}_G^t - \mathbf{P}_D^t) \le \overline{\mathbf{F}},\ (t=1,2,...,24) \qquad (3.75)$$

$$\underline{P}_{Gi}^t \le P_{Gi}^t \le \overline{P}_{Gi}^t,\ (t=1,2,...,24)\ (i \in I) \qquad (3.76)$$

The start-up/shut-down fee in the objective function is a constant, making the calculation a standard security-constrained economic dispatch problem. We will solve it to obtain the generation scheduling and nodal price at the moment t. Usually prices vary over time and this is the basis of the time-of-use electric price.

Figure 3.11 depicts a simple example. The bidding prices for units 1, 2, and 3 are $10, 20, and 30/MW, respectively, and $0 for start-up and shut-down costs. Without considering grid constraints, the electric prices are

- $10 from time period 0 to 6
- $20 from time period 7 to 8
- $30 from time period 9 to 20
- $20 at time period 21
- $10 from time period 22 to 23

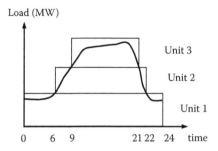

Figure 3.11   Time-of-use electric prices.

Note that prices are higher in peak hours and lower in off-peak hours. This is a positive feature of time-of-use electric price because it can guide the loads to use more power at off-peak times. Assuming that grid constraints are not considered and the start-up/shut-down bidding price is 0, we may aggregate all nodal loads for time period t and find aggregated load

$$\sum_{i=1}^{NB} P_{Di}^t$$

Equations 3.73 to 3.76 may thus be expressed as

$$\underset{P_G,U}{Min} \quad \sum_{t=1}^{24} \sum_{i=1}^{NG} U_i^t c_i P_{Gi}^t \tag{3.77}$$

$$S.T. \quad \mathbf{e}^T \mathbf{P}_G^t - \sum_{i=1}^{NB} P_{Di}^t = 0, \ (t = 1, 2, \ldots, 24) \tag{3.78}$$

$$U_i^t \underline{P}_{Gi}^t \leq P_{Gi}^t \leq U_i^t \overline{P}_{Gi}^t, \ (t = 1, 2, \ldots, 24) \ (i = 1, 2, \ldots, NG) \tag{3.79}$$

$$U_i^t \in \{0,1\}, \ (t = 1, 2, \ldots, 24) \ (i = 1, 2, \ldots, NG) \tag{3.80}$$

No coupling occurs among time periods. Thus each time period may be calculated independently. The ranking method (often called the merit order algorithm) may be used again to solve this problem. We first turn on the generating unit with the minimum bidding price (minimum c), allow it to operate at its nominal power output, and then check whether the demand is satisfied. If the answer is yes, this unit becomes the marginal generating unit and its bidding price becomes the marginal price. Otherwise, we turn on the unit with the second minimum bidding price and allow it to operate at its nominal output, and again check whether the demand is satisfied. We repeat this calculation until total load demand is satisfied.

3.6.2 Multi-Settlement System

Day-ahead markets may or may not be financially binding. When the day-ahead market is financially binding, the price of a real-time market has financial meaning only for the output difference between day-ahead and real-time markets. This is called the multi-settlement system method. For example, the output of a generating unit is 100 MW in a day-ahead market and the clearing price is $20/MW. In a real-time market, the unit may actually output 80 MW with a real-time clearing price of $25/MW. Thus, although it receives $20 × 100 = $2000 in the day-ahead market, it should return $25 × (100 − 80) = $500 in the real-time market. The multi-settlement system is similar to the contract for difference discussed in Section 3.2.2. They differ in that the contract electricity in the contract for difference arrangement may be specified annually or monthly.

Usually, generating units take day-ahead markets seriously in a multi-settlement system, making market operations stable. This can be demonstrated by an example. Assume that a generating unit wins 100 MW in a day-ahead market bid. If it output 10 MW less in real time, it would have to purchase 10 MW from the real-time market. Because the real-time price would be higher than that of the day-ahead market based on 10 MW less electricity, the generating unit would suffer economic losses. To prevent this, the unit would have to generate as much as 100 MW in real time.

Another outstanding feature of the multi-settlement system is that it allows loads to declare consumption to lock day-ahead electric prices. When a day-ahead price gets too high, loads will be able to protect themselves by decreasing their demands in the real-time market. In essence, the self-motivated demand response mechanism is an advantage of the multi-settlement market.

3.7 Ex Post Spot Pricing

The nodal prices introduced in earlier sections take effect ex ante and are based on the assumption that a load forecast is absolutely accurate and generating units are subject to dispatch instructions. They are known as ex ante prices and the related dispatch is called the ex ante dispatch. However, in a real world, these two conditions are rarely satisfied because errors in load forecasts and deviations in unit operation occur. As a result, ex ante prices are not always in line with actual system operations.

Let us consider an ideal example to explain the issue. Assume during a period without congestion one marginal generating unit has two bidding prices ($10/MW

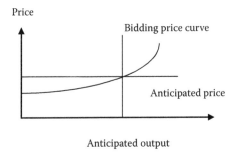

Figure 3.12   Unit's anticipated output and electric price.

and $30/MW) according to output electricity, as illustrated in Figure 3.12. Assume also that the ex ante dispatch requires this unit to operate at its high price range; the system clearing price is thus $30/MW. In practice, the unit may be operating at its low price range and the practical clearing price should not be $30/MW.

Ex post spot pricing calculates nodal marginal electric prices based on real power system operations. The basic process of ex post pricing is illustrated below. Assume a dispatch period from 12:00 to 12:05.

12:00—Load forecasting for 12:05. Perform contingency analysis, determine active transmission constraints, calculate output and price for 12:05, and broadcast prices to all transmission zones.

12:05—State estimation. Determine actual outputs and price range for 12:05, calculate ex post electric prices, and publish the prices on Internet.

We now introduce the mathematical model for ex post spot pricing. If a system operator receives bidding from generator units before dispatching, it can conduct a security-constrained economic dispatch for the following time period:

$$\underset{\mathbf{P}_G}{Min} \quad \mathbf{p}^T \mathbf{P}_G \tag{3.81}$$

$$S.T. \quad \mathbf{e}^T (\mathbf{P}_G - \mathbf{P}_D) = 0 \tag{3.82}$$

$$\mathbf{T}(\mathbf{P}_G - \mathbf{P}_D) \leq \overline{\mathbf{F}} \tag{3.83}$$

$$\underline{\mathbf{P}}_G \leq \mathbf{P}_G \leq \overline{\mathbf{P}}_G \tag{3.84}$$

Notice that only active constraints manually selected by operators are included in the optimization. Defining λ, μ, $\check{\tau}$, and $\hat{\tau}$ as the Lagrange multipliers for each constraint, we have the Lagrange function:

$$\Gamma = \mathbf{p}^T \cdot \mathbf{P}_G + \lambda \mathbf{e}^T (\mathbf{P}_G - \mathbf{P}_D) + \mu^T [\mathbf{T}(\mathbf{P}_G - \mathbf{P}_D) - \bar{\mathbf{F}}] +$$
$$\hat{\tau}^T (\mathbf{P}_G - \bar{\mathbf{P}}_G) + \check{\tau}^T (\underline{\mathbf{P}}_G - \mathbf{P}_G) \tag{3.85}$$

where the nodal price for node j is

$$\rho_j = -\lambda - \sum_{i=1}^{m} \mu_i T_{ij} \tag{3.86}$$

As stated above, the nodal price has only "anticipation" value. It is based on the assumption that the load forecast is absolutely accurate and generating outputs are completely consistent with the dispatch instructions and is thus called the ex ante price. Based on the principle of ex post spot pricing, the ex ante price is not used for financial settlement. The system operator notifies all generating units about the anticipated price via communication networks, but does not specify their outputs. Different generating units will respond differently after receiving this dispatch instruction. For example

- A generating unit outputs exactly the amount of electricity anticipated. This result indicates that the unit's bidding price equals its actual operation cost, as illustrated in Figure 3.12.
- A generating unit outputs more electricity than anticipated. This indicates that the unit's bidding price is higher than its actual operation cost, and the unit is thus willing to output more to lift the ex post price. Another possibility is that the unit's actual ramp rate is larger than what is reported to the system operator.
- A generating unit outputs less electricity than anticipated. This may indicate that its bidding price is lower than the actual operation cost and it is thus willing to output less. Note that this behavior will not be punished. Another possibility is that the unit's actual ramp rate is smaller than reported to the system operator.

The calculation of ex post electric price begins at the end of a dispatching period, when state estimation data for this period is available. We use vector variable $\hat{\mathbf{P}}$ to denote actual dispatched outputs derived from the estimated states and ε to denote a vector with a small value (i.e.. 1 MW). We first solve the following problem:

$$\underset{\mathbf{P}_G}{Min} \quad \mathbf{p}^T \cdot \mathbf{P}_G \tag{3.87}$$

$$S.T. \quad \mathbf{e}^T (\mathbf{P}_G - \hat{\mathbf{P}}_G) = 0 \tag{3.88}$$

$$\mathbf{TP}_G \leq \mathbf{T}\hat{\mathbf{P}}_G \qquad\qquad (3.89)$$

$$0 \leq \mathbf{P}_G \leq \hat{\mathbf{P}}_G + \varepsilon$$

This model is similar to general nodal pricing models. We can obtain ex post electric prices by solving the model but we must consider the following important market rules:

1. Undispatchable generating units are not allowed to participate in the above optimization and thus cannot set prices. The outputs of these units are assumed to be constants equal to their actual output values and are removed in power balance equations and load flow inequality constraints.
2. Generating units outputting 110% or more than anticipated are not allowed to participate in ex post pricing to prevent over-performance.
3. The ex post price is lower than the ex ante price. This is a soft rule. When the computed ex post price is higher than the ex ante price, system operators must analyze the calculation and post a notice on the Internet that the released price is to be confirmed.
4. To ensure grid security, system operators are authorized to require certain units to output at fixed megawatt levels. The units will be punished if they fail to follow the instruction.

Two facts of ex post spot pricing are worth reviewing. The power balance equation is different from the ex ante equation because the ex post price is calculated around operation points; the normal power balance equations cannot achieve this goal.

The reason to use the output inequality constraint with the form $0 \leq \mathbf{P}_G \leq \hat{\mathbf{P}}_G + \varepsilon$ is obvious: the ex post spot pricing model includes only dispatchable generating units that can set the clearing price.

We now use an example to show why the load flow inequality constraint of the ex post spot pricing model is different from that of a traditional dispatch model. Anticipated output and electric price are illustrated in Figure 3.13(a). Actual output and electric price are illustrated in Figure 3.13(b). Note that the actual load flow is 97 MW, while the transmission capacity is 100 MW. However, we still consider this line congested in the ex post electric price calculation according to the rule covering manually selected active constraints. This explains why we use $\mathbf{TP}_G \leq \mathbf{T}\hat{\mathbf{P}}_G$ in ex post electric price calculations instead of the original load flow constraint $\mathbf{T}(\mathbf{P}_G - \mathbf{P}_D) \leq \bar{\mathbf{F}}$.

We can see from the above calculation that the ex post spot pricing model is not sensitive to generating unit data. For example, we do not need accurate data for ramp rates or available capacities. Besides, most units are not required to follow dispatch instructions strictly. When most units output less than anticipated, the electric price will increase to stimulate them to output more, and vice versa. This is a dynamic interaction that converges to equilibrium.

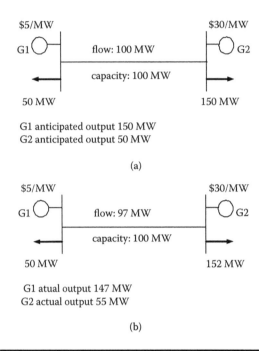

Figure 3.13 **Ex post electric price calculation. (a) Anticipated output and electric price. (b) Ex post electric price.**

The failures of some units to obey dispatch instructions will not cause serious consequences if the system capacity is large enough. Therefore, ex post spot pricing is considered a good method. It has been adopted in Pennsylvania, Maryland, New Jersey, New York, and New England in the U.S., Ontario, Canada, New Zealand, and other regions. For an excellent account on ex post pricing, the readers are referred to On Ex Post Pricing in the Real-Time Electricity Market (Zheng and Litvinov, 2011).

3.8 Transmission Losses

We did not consider transmission losses in the above pricing models. According to experience, transmission losses usually represent 3 to 5% of the total load and cannot be ignored. Thus we must consider transmission losses in the operation of a practical electricity market. As previously stated, in pool-based electricity markets, buyers and sellers trade at marginal prices so we can solve this problem by incorporating losses into marginal prices. A nodal pricing model including transmission loss is introduced below.

We first determine the relationship between nodal input power and transmission loss in a power system. Assume that $\mathbf{P}_G$ and $\mathbf{Q}_G$ are a generator's active and reactive outputs, respectively, $\mathbf{P}_D$ and $\mathbf{Q}_D$ are active and reactive parts of a load, and $\mathbf{P}(\mathbf{V}, \boldsymbol{\theta})$ and $\mathbf{Q}(\mathbf{V}, \boldsymbol{\theta})$ are nodal active and reactive power inputs that are functions of voltage amplitudes and angles. The system load flow equations in polar coordinates can be expressed as follows:

$$\mathbf{P}_G - \mathbf{P}_D = \mathbf{P}(\mathbf{V}, \boldsymbol{\theta}) \tag{3.90}$$

$$\mathbf{Q}_G - \mathbf{Q}_D = \mathbf{Q}(\mathbf{V}, \boldsymbol{\theta}) \tag{3.91}$$

From these equations, we can see an underlying relationship between nodal input power and transmission loss. If we use P_{Loss} to denote transmission loss, and $\mathbf{e}$ is a vector of ones, the system transmission loss can thus be denoted by

$$\mathbf{e}^T (\mathbf{P}_G - \mathbf{P}_D) - P_{Loss}(\mathbf{V}, \boldsymbol{\theta}) = 0 \tag{3.92}$$

We can employ the P-Q decoupled load flow method to solve the load flow equations and obtain linearly approximate solutions. Using $\Delta\boldsymbol{\theta}$ and $\Delta\mathbf{V}$ to denote corrections and $\Delta\boldsymbol{\theta}'$ to denote $\Delta\boldsymbol{\theta}$ without components relevant to the reference node, we have

$$\Delta\boldsymbol{\theta}' \approx \mathbf{B}'^{-1}(\Delta\mathbf{P}_G - \Delta\mathbf{P}_D) \tag{3.93}$$

$$\Delta\mathbf{V} \approx \mathbf{B}''^{-1}(\Delta\mathbf{Q}_G - \Delta\mathbf{Q}_D) \tag{3.94}$$

In these equations, Δ denotes a small incremental value; $\mathbf{B}'$ and $\mathbf{B}''$ are constant matrixes derived from the nodal admittance matrix that is explained in many books about power system analysis and requires no elaboration here. We can learn from Equations 3.94 and 3.95 that the angles of voltages are determined by nodal active power input, and the magnitudes are determined by nodal reactive power input. P-Q decoupled load flow equations are accurate enough for dispatching purposes and have been adopted widely. We now employ a first order Taylor series expansion to transmission loss at one operation point and obtain:

$$P_{Loss}(\mathbf{V}, \boldsymbol{\theta}) = \hat{P}_{Loss} + \frac{\partial P_{Loss}}{\partial \mathbf{V}}^T \Delta\mathbf{V} + \frac{\partial P_{Loss}}{\partial \boldsymbol{\theta}}^T \Delta\boldsymbol{\theta} \tag{3.95}$$

Here we have $\hat{P}_{Loss}$ as the transmission loss at the Taylor series expansion point. Since

$$P_{Loss} = \sum_{k \in all\_lines} G_k \left[V_i^2 + V_j^2 - 2V_i V_j \cos(\theta_i - \theta_j) \right] \tag{3.96}$$

By further assuming that $V_i \approx 1$, $V_j \approx 1$, and $\theta_i - \theta_j$ is very small, we have

$$P_{Loss} \approx \sum_{k \in all\_lines} G_{ij}(\theta_i - \theta_j)^2 \tag{3.97}$$

Thus we easily obtain $\partial P_{Loss}/\partial \theta$. If we further assume that the voltage of a generator (PV) node is fixed during a dispatching cycle, we have $\Delta V = 0$. For convenience, we assume the last node to be the reference node:

$$\Delta \mathbf{P}_G = \begin{bmatrix} \Delta \mathbf{P}'_G \\ \Delta P_{GN} \end{bmatrix} \tag{3.98}$$

$$\Delta \mathbf{P}_D = \begin{bmatrix} \Delta \mathbf{P}'_D \\ \Delta P_{DN} \end{bmatrix} \tag{3.99}$$

Applying $\Delta V = 0$ to a Taylor series expansion for transmission loss according to Equation 3.97, we have

$$
\begin{aligned}
P_{Loss}(\mathbf{V}, \boldsymbol{\theta}) &= \hat{P}_{Loss} + \left(\frac{\partial P_{Loss}}{\partial \boldsymbol{\theta}} \right)^T \Delta \boldsymbol{\theta} \\
&= \hat{P}_{Loss} + \left(\frac{\partial P_{Loss}}{\partial \boldsymbol{\theta}} \right)^T \begin{bmatrix} \Delta \boldsymbol{\theta}' \\ 0 \end{bmatrix} \\
&= \hat{P}_{Loss} + \left(\frac{\partial P_{Loss}}{\partial \boldsymbol{\theta}} \right)^T \begin{bmatrix} \mathbf{B}'^{-1}(\Delta \mathbf{P}'_G - \Delta \mathbf{P}'_D) \\ 0 \end{bmatrix} \\
&= \hat{P}_{Loss} + \left(\frac{\partial P_{Loss}}{\partial \boldsymbol{\theta}} \right)^T \begin{bmatrix} \mathbf{B}'^{-1} & 0 \\ 0 & 0 \end{bmatrix} (\Delta \mathbf{P}_G - \Delta \mathbf{P}_D) \\
&\approx \hat{P}_{Loss} + \left(\frac{\partial P_{Loss}}{\partial \mathbf{P}_G} \right)^T (\Delta \mathbf{P}_G - \Delta \mathbf{P}_D)
\end{aligned}
\tag{3.100}
$$

where

$$\left(\frac{\partial P_{Loss}}{\partial \mathbf{P}_G} \right)^T = \left(\frac{\partial P_{Loss}}{\partial \boldsymbol{\theta}} \right)^T \begin{bmatrix} \mathbf{B}'^{-1} & 0 \\ 0 & 0 \end{bmatrix} \tag{3.101}$$

Substituting the above equation into the power balance equation, we have

$$\left(\mathbf{e} - \frac{\partial P_{Loss}}{\partial \mathbf{P}_G} \right)^T (\mathbf{P}_G - \mathbf{P}_D) = \hat{P}_{Loss} - \left(\frac{\partial P_{Loss}}{\partial \mathbf{P}_G} \right)^T (\hat{\mathbf{P}}_G - \hat{\mathbf{P}}_D)$$

$$= \mathbf{e}^T \left(\hat{\mathbf{P}}_G - \hat{\mathbf{P}}_D \right) - \left(\frac{\partial P_{Loss}}{\partial \mathbf{P}_G} \right)^T \left(\hat{\mathbf{P}}_G - \hat{\mathbf{P}}_D \right) \tag{3.102}$$

Assuming $\boldsymbol{\beta} = \mathbf{e} - \dfrac{\partial P_{Loss}}{\partial \mathbf{P}_G}$, the system power balance constraint may be expressed as

$$\boldsymbol{\beta}^T (\mathbf{P}_G - \mathbf{P}_D) = \boldsymbol{\beta}^T (\hat{\mathbf{P}}_G - \hat{\mathbf{P}}_D) \tag{3.103}$$

Equation 3.103 established a connection between transmission losses of two close operation points. The real-time dispatching model with consideration of transmission loss is thus expressed by

$$\underset{\mathbf{P}_G}{Min} \qquad \mathbf{p}^T \mathbf{P}_G \tag{3.104}$$

$$S.T. \qquad \boldsymbol{\beta}^T (\mathbf{P}_G - \mathbf{P}_D) = \boldsymbol{\beta}^T (\hat{\mathbf{P}}_G - \hat{\mathbf{P}}_D) \tag{3.105}$$

$$\mathbf{T}(\mathbf{P}_G - \mathbf{P}_D) \le \overline{\mathbf{F}} \tag{3.106}$$

$$\underline{\mathbf{P}}_G \le \mathbf{P}_G \le \overline{\mathbf{P}}_G \tag{3.107}$$

The Lagrange function of this optimization is

$$\Gamma = \mathbf{p}^T \mathbf{P}_G + \lambda \boldsymbol{\beta}^T [(\mathbf{P}_D - \mathbf{P}_G) - (\hat{\mathbf{P}}_D - \hat{\mathbf{P}}_G)] + \boldsymbol{\mu}^T [\mathbf{T}(\mathbf{P}_G - \mathbf{P}_D) - \overline{\mathbf{F}}] +$$
$$\check{\boldsymbol{\tau}}^T (\underline{\mathbf{P}}_G - \mathbf{P}_G) + \hat{\boldsymbol{\tau}}^T (\mathbf{P}_G - \overline{\mathbf{P}}_G) \tag{3.108}$$

According to Karush–Kuhn–Tucker conditions:

$$\frac{\partial \Gamma}{\partial \mathbf{P}_G} = \mathbf{p} - \check{\boldsymbol{\tau}} + \hat{\boldsymbol{\tau}} - \lambda \boldsymbol{\beta} - \mathbf{T}^T \boldsymbol{\mu} = 0 \tag{3.109}$$

and the nodal electric price is expressed by

$$\boldsymbol{\rho} = \lambda \boldsymbol{\beta} + \mathbf{T}^T \boldsymbol{\mu} = \lambda \mathbf{e} - \lambda \frac{\partial \mathbf{P}_{Loss}}{\partial \mathbf{P}_G} + \mathbf{T}^T \boldsymbol{\mu} \tag{3.110}$$

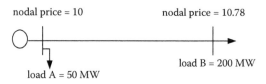

Figure 3.14 Nodal price with consideration of transmission loss.

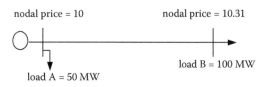

Figure 3.15 Nodal price changes as operation point varies.

From Equation 3.110, we can clearly see the three components of nodal electric price: electric power, transmission loss, and congestion.

Generally, the fees charged from loads are higher than those paid to generation companies and the difference is called the merchandise surplus—a combination of transmission loss merchandise surplus and congestion merchandise surplus. In some markets in the U.S. and elsewhere that employ nodal electric pricing, congestion merchandise surplus is allocated to loads via financial transmission rights. Transmission loss merchandise surplus is usually proportionally allocated to loads and the result is sometimes unfair. We use a two-node example (Figure 3.14) to demonstrate this. The figure shows that transmission loss merchandise surplus should be completely allocated to load B and not allocated based on the sizes of loads A and B. The transmission loss problem for power systems with loops has not been solved.

Another problem related to transmission loss is that the loss factor varies with operation point. Figure 3.15 shows that when a system operation point changes, a unit's bidding price remains the same but the nodal electric price changes. In practice, the real-time calculation of loss factors is impossible and the lack of loss factor data affects the fairness of a dispatch algorithm to a degree. A compromise is to use the average loss factors of several typical operation situations. This method was employed by the East China power grid starting in 2004.

Dealing with transmission losses can be much more complicated in a bilateral trade market. Practical electricity markets often involve both pool and bilateral trading, which further complicates the issue. We will not pursue this topic further in this chapter.

3.9 Bilateral Trading in the United Kingdom

In the United Kingdom, England and Wales share a grid and Scotland and Northern Ireland have their own grids. Total installed capacity as of 2003 was 78,524 MW. Before privatization in 1990, power systems in the U.K. operated as an integrated monopoly. Since then, the U.K. power industry has undergone three periods of marketization.

The first marketization, known as the power pool period, featured a pool-based operational mode. The second was marked by the implementation of the New Electricity Trading Arrangement (NETA) and involved bilateral trading between generation companies and customers. The NETA period started in March 2001. The third marketization implemented the British Electricity Trading and Transmission Arrangements (BETTA) characterized by unified management of the whole British power system by a single corporation. The BETTA period started in 2005.

The U.K. has been studying and implementing bilateral trading since 1997 and officially entered the bilateral market in March 2001. The market has not encountered major problems to date. The establishment of a British bilateral trading market was a milestone in electricity market history or, more exactly, power system operation history. This section focuses on this bilateral market. The ERCOT electricity market in Texas has much in common with the British market.

3.9.1 Motivation

From 1990 to 2001, the British pool-based electricity market introduced bidding mechanisms on the generation side that allowed independent generation companies to bid and sell electricity to power pools. Distribution companies and electric suppliers bought electricity from pools and then sold it to customers. Generation companies with capacities exceeding 10 MW and electric suppliers with capacities above 50 kW were power pool participants. Customers could choose electric suppliers but not distribution companies.

In a pool-based electricity market, the demand side can only accept clearing prices created through generation companies' bidding processes passively. The demand side has no means to influence final price. Many believe that such markets are incomplete because only sellers have market power. To prevent this and build a complete market, the demand side should be allowed to participate in pricing.

As a contrast to a centralized market under uniform management of a system operator, a bilateral market can increase pressure on generation companies and load responses. Moreover, pool-based markets present disadvantages such as complicated structures, low market transparency, and high cost of software development and maintenance.

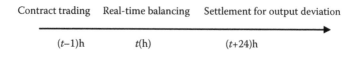

Figure 3.16   Schedule for market activities.

3.9.2 Structure and Mechanism

In a bilateral trading electricity market, sellers and buyers may freely sign contracts for electricity supplies and consumption. The U.K. market designers believe that the contract and trading modes of most commodities are applicable to electricity markets. The contracts usually include certain elements such as trading period, size, and price. Contract-related cash flows are also determined by both sides and have nothing to do with system operators.[*]

The British electricity market began to use the bilateral trading mode in 2001. The real-time property and transmission capacity limits of power systems create several problems in establishing a bilateral electricity market, for example (1) how to maintain real-time electricity balance (balancing mechanism) and (2) how to curtail generation contracts to satisfy transmission capacity limits (congestion management).

Common sense dictates that supply and demand sides of a contract are usually willing to allow system operators to correct contracts as long as the supply and demand sides receive compensation. Based on this principle, one method to deal with balancing and congestion problems is to employ a bidding and tendering mechanism known as compensation for changed contract bidding. Figure 3.16 illustrates a schedule for market activities under this mechanism.

3.9.3 Real-Time Balancing

Generating units should notify the system operator of their contract electricity for every trading period (30 minutes in the U.K.) by 11:00 a.m. one day ahead. The process is called initial physical notification. Moreover, notification for final contract electricity for one trading period should be submitted 1 hour before this trading period. Contract electricity output curves and load curves provided by units should be accurate on a minute scale. The submission of final contract electricity output data is designated the final physical notification (FPN) and is illustrated in Figure 3.17.

After contracts are submitted, we must arrive at a balancing market that allows a system operator to decide whether to increase or decrease unit outputs or loads

---

[*] Some thought that prices stated in agreements should be open to public review to encourage competition.

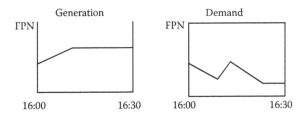

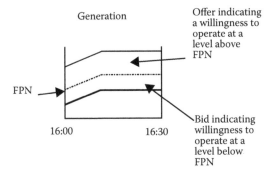

Figure 3.17 Final physical notification in balance market. (From R. Haigh and J. Bagwell. 2000. An Overview of the New Electricity Trading Arrangements. http://www.ofgem.gov.uk)

Figure 3.18 Quotation price of generating unit in balance market. (From R. Haigh and J. Bagwell. 2000. An Overview of the New Electricity Trading Arrangements. http://www.ofgem.gov.uk)

based on final contracts in order to maintain system balance. The bid and offer data provided by units or loads consists of a series of bid and offer data pairs.

An offer price is the quotation price for increasing a unit's output or decreasing its load. The bid price indicates the quotation price for decreasing a unit's output or increasing its load. Figure 3.18 illustrates the quotation price of a generating unit in a balance market. Note that the offer price is usually higher than the bid price for a generating unit. The settlement rule in a NETA balancing market is "pay as bid" (PAB). This is an important but controversial feature of the British electricity market.

We now introduce the balance dispatch model. We have not included load participation in order to maintain simplicity. For every single dispatching period (i.e., 5 minutes), a system operator balances the market based on generating units' bidding prices. We may assume the units' contract outputs and loads are constant during this short period. We designate contract output as P_G^0 and contract load as P_D. We denote generating units' upward and downward outputs required by the system operator as ΔP_G^+ and ΔP_G^- and relative bidding prices as

p^+ and p^-. Thus the system operator's job is to work out the $\Delta \mathbf{P}_G^+$ and $\Delta \mathbf{P}_G^-$ pair to minimize total balance cost

$$\sum_{i=1}^{NG} \left(p_i^+ \Delta P_G^+ + p_i^- \Delta P_G^- \right)$$

with load $\mathbf{P}_D$ and grid constraints satisfied. Note that $(\mathbf{P}_G^0 + \Delta \mathbf{P}_G^+ - \Delta \mathbf{P}_G^-)$ is the vector of a unit's actual outputs. The problem can thus be expressed as

$$\underset{\Delta \mathbf{P}_G^+, \Delta \mathbf{P}_G^-}{Min} \sum_{i=1}^{NG} \left(p_i^+ \Delta P_{Gi}^+ + p_i^- \Delta P_{Gi}^- \right) \tag{3.111}$$

$$S.T. \quad \mathbf{e}^T \left(\mathbf{P}_G^0 + \Delta \mathbf{P}_G^+ - \Delta \mathbf{P}_G^- - \mathbf{P}_D \right) = 0 \tag{3.112}$$

$$\mathbf{T}(\mathbf{P}_G^0 + \Delta \mathbf{P}_G^+ - \Delta \mathbf{P}_G^- - \mathbf{P}_D) \le \overline{\mathbf{F}} \tag{3.113}$$

$$\underline{\mathbf{P}}_G \le \mathbf{P}_G^0 + \Delta \mathbf{P}_G^+ - \Delta \mathbf{P}_G^- \le \overline{\mathbf{P}}_G \tag{3.114}$$

$$\Delta \mathbf{P}_G^+ \ge 0, \ \Delta \mathbf{P}_G^- \ge 0 \tag{3.115}$$

This problem can be solved by a standard linear programming algorithm. The model used by the ERCOT market in the U.S. since 2005 is similar to the British one.

Figure 3.19 illustrates a 30-minute balance dispatching period during which the bid and offer price of a generating unit is accepted. Its output is first increased and part of its electricity in Offer 1 and Offer 2 is purchased at £30/MWh and £40/MWh. Soon afterward, some of its output in Bid 1 is decreased and the system operator pays £18/MWh.

Another important feature of the British bilateral trading electricity market is that generation scheduling is conducted by power plants independently. Thus, the technical support system of system operators does not include software for generation scheduling. This greatly simplified the system. However, we must note that for power systems with low unit start-up speeds, the British balancing market design may not be applicable.

3.9.4 Performance

Here we show the performance of the market with two prices: (1) the system buy price (SBP) derived from weighted offer prices; and (2) the system sell price (SSP)

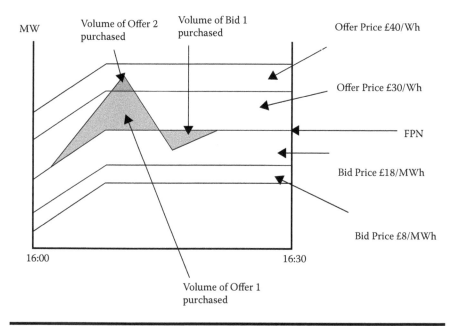

Figure 3.19 Dispatching in a balance market. (From R. Haigh and J. Bagwell. 2000. An Overview of the New Electricity Trading Arrangements. http://www. ofgem.gov.uk)

derived from weighted bid prices. We know from experience that the SBP is generally higher and varies greatly while the SSP is lower and varies less, as shown in Figure 3.20. As the figure indicates, the difference of daily average SBP and SSP decreased from £70/MWh to £17/MWh. Figure 3.21 shows a gradual decrease of balance market cost. The two figures indicate the maturity of the electricity market.

The British bilateral trading electricity market started in March 2001. At present, 97% of electricity trading is conducted through bilateral contracts and only 2 to 3% is traded through a balancing market. In the first year, the electric price dropped a lot, proving reform was successful, as shown in Figure 3.22.

3.10 Electricity Market Reform in California

California has been the pioneer of power industry reforms in the United States. In 1996, its legislature passed Assembly Bill 1890 to introduce retail electricity competition and promote power industry reform and established two independent organizations (PX and ISO) to handle market business and power grid security separately. PX is in charge of most market trading. ISO is responsible for system operations and tries not to disturb market operations. On March 1, 1998, the reformed

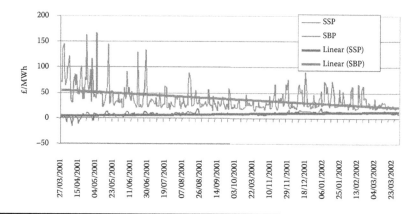

Figure 3.20 Daily average SSP and SBP after start-up of bilateral trading market. (From R. Haigh and J. Bagwell. 2000. An Overview of the New Electricity Trading Arrangements. http://www.ofgem.gov.uk)

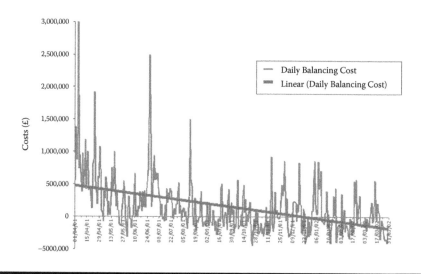

Figure 3.21 Balance cost. (From R. Haigh and J. Bagwell. 2000. An Overview of the New Electricity Trading Arrangements. http://www.ofgem.gov.uk)

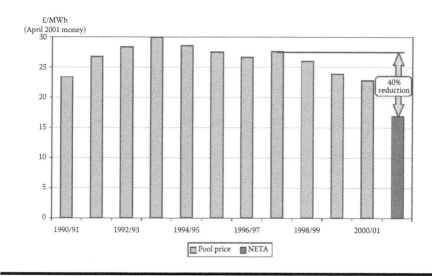

Figure 3.22 Pool price and NETA price. (From Office of Gas and Electricity Markets. 2002. Review of the First Year of NETA. http://www.ofgem.gov.uk)

Californian electricity market, aiming at promoting competition and economic efficiency, started official operation and PX and ISO began working. The three major power suppliers in California are required to trade in PX. Three electric utilities sold most of their thermal generation assets.

However, the Californian electricity market encountered an unprecedented electric power crisis just 2 years later. In December 2000, the wholesale electric price reached $400/MWh, and the daily loss of electric utilities totaled about $50 million. By spring of 2001, the two major electric utilities accumulated total debts of $20 billion and partial power cuts started. To understand the causes for this crisis, let us first examine the external reasons:

■ California gas prices kept rising from May 2000 to the end of the year—about five times the increases in other states. Since thermal generating units were usually marginal, gas prices affected electric prices.

■ High temperatures in the summer of 2000 caused the total load to rise while the retail price remained regulated.

■ California relied on electricity imported from other states to compensate for its insufficient generation capacity. In the summer of 2000, a drought in the northwestern U.S. greatly reduced imported electricity.

■ Many generating units were forced to shut down due to emission pollution.

■ Many generating units shut down for maintenance in the fall after handling high loads in the summer.

Analyzing the crisis based on market mechanisms, we can see additional causes:

- Generation ownership was too concentrated.
- The wholesale market did not consider demand side responses.
- The market lacked long-term contracts. The three major electric utilities were not permitted to sign bilateral futures contracts and were urged to buy electricity from the PX day-ahead market instead.

On January 30, 2001, the Californian power exchange center was closed and nodal pricing began to emerge. The current Californian electricity market has the following features:

- Buyers and sellers trade mainly via bilateral contracts that constitute 95% of total electricity wholesale market.
- Distributed generation scheduling was adopted to allow generation companies to schedule independently to fulfill their contracts.
- Congestion between different regions is solved by a regional pricing model.
- A special congestion management system ensures the balance of generation and load for every scheduling coordinator (SC).

Congestion management requires explanation. Before the daily market opens, the California system operator requires every SC to provide a balanced generation load schedule and unit bidding prices. The parameters of schedules are derived from day-ahead bilateral trading. As an example, assume the number of market participants is N_{SC}. To eliminate congestion or real-time imbalance, the California system operator employed the following mathematical optimization model:

$$\underset{\mathbf{P}_G}{Min} \quad \mathbf{p}^T \mathbf{P}_G \tag{3.116}$$

$$S.T. \quad \mathbf{e}^T (\mathbf{P}_G - \mathbf{P}_D) = 0 \tag{3.117}$$

$$\mathbf{T}(\mathbf{P}_G - \mathbf{P}_D) \le \overline{\mathbf{F}} \tag{3.118}$$

$$\underline{\mathbf{P}}_G \le \mathbf{P}_G \le \overline{\mathbf{P}}_G \tag{3.119}$$

$$\sum_{i=1}^{NG} P_{Gk,i} = \sum_{i=1}^{NB} P_{Dk,i}, \ (k = 1, 2, \ldots, N_{SC}) \tag{3.120}$$

The model is unique because of Equation 3.120, also known as the market decoupling constraint. This special constraint is intended to reduce the system operator disturbances of the market. If ISO needs to modify the energy plan of an SC to help deal with congestion, the redispatch should not break the supply–demand balance in the plan.

References

R. Haigh and J. Bagwell. 2000. An Overview of the New Electricity Trading Arrangements. http://www.ofgem.gov.uk

D. Luenberger, 1995. *Microeconomic Theory*. New York: McGraw Hill.

Office of Gas and Electricity Markets. 2002. Review of the First Year of NETA. http://www.ofgem.gov.uk

F. C. Schweppe, M. C. Caramanis, R. D. Tabors et al. 1988. *Spot Pricing for Electricity*. Boston: Kluwer.

T. Zheng, E. Litvinov. 2011. On Ex Post Pricing in the Real-Time Electricity Market. *IEE Trans. on Power Systems*, vol. 25, no. 1, pp. 153–165.

Bibliography

D. Feng, Z. Xu, J. Østergaard et al. 2012. Spot pricing when Lagrange multipliers are not unique. *IEEE Transactions on Power Systems*, 27, 314–322.

R. J. Green. 2000. Competition in generation: the economic foundations. *Proceedings of IEEE*, 88, 128–139.

Chapter 4

Market Design: Procurement of Ancillary Services

Reserves and automatic generation control (AGC) are indispensable for power system operation. However, the costs of providing these ancillary services are not covered in active power markets. Therefore, we must find ways to compensate for these costs that consist of equipment sunk cost, operation and maintenance cost, or opportunity cost for providing energy and other services. In this section, we start with an introduction of the market-based reserve procurement model, then extend the model for AGC procurement, and finally combine them into an energy market. A compensation mechanism that does not rely on market solutions is also described. This compensation mechanism was originally adopted in Australia and later in the Zhejiang market in China and is discussed as an alternative.

4.1 Reserve Market

The principle for reserve market design is that capacity compensation fees should be paid based on market clearing prices for generators that have the ability to provide reserves, that is, payments should be made to generators that provide reserve capacity through market competition. In this mechanism, the generators in a system are expected to bid the lowest prices for providing their reserve capacities. Generally, the dispatch center purchases reserve capacity for each dispatch period (for example, every 5 minutes) and the costs for reserve capacity are covered by customers. Let us look at a simple example. In a two-generator market consisting of generators A and B (Gan and Litvinov 2003), the energy-only dispatch problem is as follows:

$$Min \quad 10P_{GA} + 20P_{GB} \tag{4.1}$$

$$S.T. \quad P_{GA} + P_{GB} = 120 \tag{4.2}$$

$$0 \le P_{GA} \le 100 \tag{4.3}$$

$$0 \le P_{GB} \le 100 \tag{4.4}$$

where P_{GA} and P_{GB} are energy outputs of units A and B, respectively; the coefficients 10 and 20 in the objective function are the bid prices of energy. Obviously, the optimal solution to this optimization problem is $\hat{P}_{GA} = 100$, $\hat{P}_{GB} = 20$, the energy clearing price is 20, the total energy cost is $10P_{GA} + 20P_{GB} = 1400$, the total income of generators (energy clearing price × power output) is 2400, and the total payment of loads is 2400. Suppose the reserve requirement is 20, and we want to dispatch energy and reserve to minimize energy and reserve production costs. The dispatch problem is then

$$Min \quad 10P_{GA} + 20P_{GB} + 30R_A + 15R_B \tag{4.5}$$

$$S.T. \quad P_{GA} + P_{GB} = 120, \, R_A + R_B = 20 \tag{4.6}$$

$$P_{GA} + R_A \le 100, \, P_{GB} + R_B \le 100 \tag{4.7}$$

$$P_{GA} \ge 0, \, 999999 \ge R_A \ge 0 \tag{4.8}$$

$$P_{GB} \ge 0, \, 999999 \ge R_B \ge 0 \tag{4.9}$$

where, R_A and R_B are the reserve dispatches and the coefficients 30 and 15 in the objective function are the reserve bids. The optimal solution of this optimization problem is found easily as $P_{GA} = 100$, $P_{GB} = 20$, $R_A = 0$, $R_B = 20$. The marginal prices for energy and reserve are 20 and 15, respectively, as will be shown later in this section. This method is an example of co-optimization of energy and reserve. The idea of co-optimization is to solve the energy and reserve dispatch in a unified way. The dispatch problem for energy and reserve can be formulated as

$$\underset{\mathbf{P}_G,\mathbf{R}}{Min} \quad \mathbf{p}^T\mathbf{P}_G + \mathbf{r}^T\mathbf{R} \tag{4.10}$$

$$S.T. \quad \mathbf{e}^T\mathbf{P}_G - P_D = 0 \tag{4.11}$$

$$\mathbf{e}^T\mathbf{R} - D_R = 0 \tag{4.12}$$

$$\underline{\mathbf{P}}_G \leq \mathbf{P}_G, \ \mathbf{P}_G + \mathbf{R} \leq \overline{\mathbf{P}}_G \tag{4.13}$$

$$0 \leq \mathbf{R} \leq \overline{\mathbf{R}} \tag{4.14}$$

where $\mathbf{R}$ is the vector for generators' reserve outputs, $\mathbf{r}$ is the vector for generators' reserve bids, $\overline{\mathbf{R}}$ is the vector for generators' ramping limits (if the dispatch period is 10 minutes, then $\overline{\mathbf{R}} = 10\mathbf{v}_R$), and D_R represents the system's reserve requirement. Now let us look at the solution method for the co-optimization problem. Let $\check{\tau}$ and $\hat{\tau}$ represent the Lagrange multipliers for power output constraints and $\check{\upsilon}$ and $\hat{\upsilon}$ denote the Lagrange multipliers for ramp limit constraints. The Lagrange function can be constructed as

$$\Gamma = \mathbf{p}^T\mathbf{P}_G + \mathbf{r}^T\mathbf{R} + \lambda(\mathbf{e}^T\mathbf{P}_G - P_D) + \varphi(\mathbf{e}^T\mathbf{R} - D_R) + \check{\tau}^T(\underline{\mathbf{P}}_G - \mathbf{P}_G) +$$
$$\hat{\tau}^T(\mathbf{P}_G + \mathbf{R} - \overline{\mathbf{P}}_G) - \check{\upsilon}^T\mathbf{R} + \hat{\upsilon}^T(\mathbf{R} - \overline{\mathbf{R}}) \tag{4.15}$$

According to standard Kuhn–Tucker optimal conditions, we have

$$\frac{\partial\Gamma}{\partial P_{Gi}} = p_i + \lambda - \check{\tau}_i + \hat{\tau}_i = 0, \qquad (i = 1, 2, \ldots, N_G) \tag{4.16}$$

$$\frac{\partial\Gamma}{\partial R_i} = r_i + \varphi - \check{\upsilon}_i + \hat{\upsilon}_i + \hat{\tau}_i = 0, \qquad (i = 1, 2, \ldots, N_G) \tag{4.17}$$

According to the principles of market clearing, the clearing prices for energy and reserves are $-\lambda$ and $-\varphi$, respectively. Let us subtract the first equation with the second one, we obtain a relationship between the clearing prices of the energy and reserves:

$$-\varphi = r_i - \lambda - p_i - \check{\upsilon}_i + \hat{\upsilon}_i + \check{\tau}_i \tag{4.18}$$

Suppose $R_i > 0$, according to the Kuhn–Tucker optimality condition, $\breve{\upsilon}_i = 0$, therefore,

$$-\varphi = r_i - \lambda - p_i + \hat{\upsilon}_i + \breve{\tau}_i > r_i - \lambda - p_i = r_i + \ell_i \qquad (4.19)$$

In the above formula $\ell_i = -\lambda - p_i$ represents the energy clearing price minus the energy bid price and may be viewed as an opportunity cost of reserve. This formula shows that if the above pricing principle is employed in an energy and reserve market, opportunity cost will be compensated automatically. To further illustrate this point, consider another example. Using the two-generator example system introduced earlier, assume the bid price of generator A is 0; then the dispatch problem is

$$Min \qquad 10P_{GA} + 20P_{GB} + 0R_A + 15R_B \qquad (4.20)$$

$$S.T. \qquad P_{GA} + P_{GB} = 120, R_A + R_B = 20 \qquad (4.21)$$

$$P_{GA} + R_A \le 100, P_{GB} + R_B \le 100 \qquad (4.22)$$

$$P_{GA} \ge 0, 999999 \ge R_A \ge 0 \qquad (4.23)$$

$$P_{GB} \ge 0, 999999 \ge R_B \ge 0 \qquad (4.24)$$

The solution of the above problem is $P_A = 80$, $P_B = 40$, $R_A = 20$, $R_B = 0$; energy price = 20 and reserve price = 10. We can see from the results that although generator A is the last one accepted to provide reserve service, its bid price is not equal to the market clearing price. The reserve clearing price is equal to the bid price plus the opportunity cost.

If we combine the energy market, reserve procurement, and AGC procurement, we will incur an enormous computational burden. Furthermore, the pricing theory is not fully understood yet. In practice, sequential optimization is generally adopted to clear reserve markets and AGC markets. We demonstrated the application of sequential optimization in reserve and energy markets through the above example. To apply sequential optimization in dispatch problems, we omit the reserve issue in the first step and then solve the energy-only dispatch:

$$Min \qquad 10P_{GA} + 20P_{GB} \qquad (4.25)$$

$$S.T. \qquad P_{GA} + P_{GB} = 120 \qquad (4.26)$$

$$P_{GA} \le 100, P_{GB} \le 100 \qquad (4.27)$$

$$P_{GA} \ge 0, P_{GB} \le 0 \qquad (4.28)$$

The optimal solution is $\hat{P}_{GA} = 100$, $\hat{P}_{GB} = 20$. The energy clearing price = 20. In the second step, we solve the problem of reserve dispatch. Obviously the optimal solution is $R_B = 20$.

This method is simple and easy to implement. The drawback is that the price signal is poor. As an example, the price for energy can be low when the cheapest generators are adopted first. However, the reserve price exceeds energy cost because expensive generators are used in a reserve market. Obviously, this is not what we want in a market.

4.2 AGC Market

An AGC market is more complex than energy and reserve markets because the regulation of AGC can go upward or downward. Furthermore, the dispatch decision of an AGC generator has an integer property that further complicates the problem. We will examine a model for the AGC market. Suppose I constitutes the index set of on-line generators. For generator $i (i \in I)$, let $w_i = 1$ denote the provision of AGC service and $w_i = 0$ indicates otherwise. For a dispatch period, the task is to determine vector **w** such that the demands of AGC up-regulation and down-regulation are met. The objection function of this problem is

$$\underset{P_G, A, w}{Min} \quad \sum_i p_i P_{Gi} + \sum_i w_i a_i \left(A_i^+ + A_i^- \right) \qquad (4.29)$$

For the real-time dispatch of energy and AGC, some operation constraints such as AGC capacity constraint and AGC ramp limit constraint should be considered. Then the dispatch problem of AGC and energy market can be formulated as

$$\underset{P_G, A, w}{Min} \quad \sum_i p_i P_{Gi} + \sum_i w_i a_i \left(A_i^+ + A_i^- \right) \qquad (4.30)$$

$$S.T. \quad \sum_i P_{Gi} = P_D \qquad \text{(active power balance constraint)} \qquad (4.31)$$

$$\sum_i A_i^+ \geq D_A \qquad \text{(AGC up-regulation capacity constraint)} \qquad (4.32)$$

$$\sum_i A_i^- \geq D_A \qquad \text{(AGC down-regulation capacity constraint)} \qquad (4.33)$$

$$0 \leq A_i^+ \leq 10 v_{Ai} w_i, \, i \in I \quad \text{(AGC up-regulation speed constraint)} \qquad (4.34)$$

$$0 \le A_i^- \le 10 v_{Ai} w_i, \; i \in I \quad \text{(AGC down-regulation speed constraint)} \quad (4.35)$$

$$A_i^+ + P_{Gi} \le \bar{A}_i, \; i \in I \qquad \text{(up limit of AGC and energy)} \quad (4.36)$$

$$A_i^- + \underline{A}_i \le P_{Gi}, \; i \in I \qquad \text{(down limit of AGC and energy)} \quad (4.37)$$

$$w_i \in \{0,1\}, \; i \in I,$$

Continuous variables $\mathbf{P}_G$, $\mathbf{A}^+$, and $\mathbf{A}^-$ along with the integer variable $\mathbf{w}$ appear in the above optimization problem. Thus in theory we can use 0–1 mixed-integer programming to solve the problem (Rau 1999). Another method for solving the problem is sequential optimization. We follow this approach and first solve the energy-only optimization problem:

$$\underset{P_G}{Min} \quad \sum_i p_i P_{Gi} \qquad\qquad (4.38)$$

$$S.T. \quad \sum_i P_{Gi} = P_D \qquad\qquad (4.39)$$

$$\underline{P}_{Gi} \le P_{Gi} \le \bar{P}_{Gi}, \; i \in I \qquad\qquad (4.40)$$

Suppose the optimal solution is P_{Gi}^*, $i \in I$. Then the co-optimization problem can be simplified:

$$\underset{A,w}{Min} \quad \sum_i w_i a_i \left(A_i^+ + A_i^- \right) \qquad\qquad (4.41)$$

$$S.T. \quad \sum_i A_i^+ \ge D_A, \; \sum_i A_i^- \ge D_A \qquad\qquad (4.42)$$

$$0 \le A_i^+ \le w_i * \min\left\{ 0 v_{Ai}, \bar{A}_i - P_{Gi}^* \right\}, \; i \in I \qquad (4.43)$$

$$0 \le A_i^- \le w_i * \min\left\{ 0 v_{Ai}, P_{Gi}^* - \underline{A}_i \right\}, \; i \in I \qquad (4.44)$$

$$w_i \in \{0,1\}, \; i \in I$$

Note that in the above optimization problem, $\min\{0\nu_{Ai}, \overline{A_i} - P^*_{Gi}\}$ and $\min\{0\nu_{Ai}, P^*_{Gi} - \underline{A_i}\}$ are constant because P^*_{Gi}, $i \in I$ are known. Therefore, this problem can be solved easily. As noted earlier, the bid price of the last accepted AGC generator is taken as the market clearing price in the dispatch period for the AGC market. This sequential market clearing method is simple and easy to implement. The drawback of this method is that prices obtained are not proper. In some markets, the AGC is cleared first, followed by the energy market. However, this approach still does not solve the flawed price problem.

We now consider a simple example of a five-generator energy and AGC market. The bid prices of the generators are shown in Table 4.1. Assume the bid prices for down- and up-regulation are the same and the AGC regulation limit equals the maximum power output, i.e., $\overline{A} = P^{max}_G$.

Suppose the forecast system load P_D of the trading period is 1000 MW and the expected demand D_A for AGC capacity is 40 MW. If the sequential optimization method is employed, the optimal dispatch is $w = (1,1,0,0,1)$. The power outputs of the five generators are shown in Table 4.2.

Table 4.1 AGC Parameters and Bid Prices of Generators

Generator No.	$(P^{max}_{Gi}, P^{min}_{Gi})$ (MW)	$(\overline{A^i}, \underline{A_i})$ (MW)	$10\nu_{Ai}$ (MW)	p_i ($/MW)	a^+_i (a^-_i) ($/MW)
1	(600, 300)	(600, 500)	60	150	20
2	(300, 100)	(300, 180)	30	180	25
3	(205, 80)	(205, 130)	40	200	30
4	(135, 50)	(135, 80)	27	250	18
5	(130, 50)	(130, 70)	26	300	15

Table 4.2 AGC Regulation Capacity and Power Outputs of Generators by Sequential Optimization

Generator No.	P_{Gi} (MW)	A^+_i (MW)	A^-_i (MW)
1	600	0	40
2	200	14	0
3	80	0	0
4	50	0	0
5	70	26	0

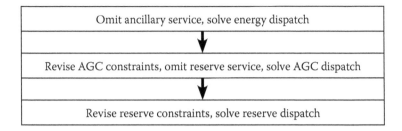

Figure 4.1 Flowchart for dispatch using sequential optimization.

4.3 Energy, Reserve, and AGC Co-Optimization Market

By combining the reserve and AGC market concepts introduced above, we can devise a model for the energy, reserve, and AGC co-optimization markets. The dispatch problem can be solved using the standard branch-and-bound or sequential optimization method. In practical systems, the algorithms used for dispatch are generally based on a certain sequential optimization method based on the Figure 4.1 flowchart. The mathematical model is as follows:

$$\underset{P_G,A,R}{Min} \quad \sum_i p_i P_{Gi} + \sum_i w_i a_i \left(A_i^+ + A_i^- \right) + \sum_i r_i R_i \tag{4.45}$$

$$S.T. \quad \sum_i P_{Gi} = P_D \qquad \text{(active power balance constraint)} \tag{4.46}$$

$$\mathbf{e}^T \mathbf{R} - D_R = 0 \qquad \text{(reserve demand constraint)} \tag{4.47}$$

$$\sum_i A_i^+ \ge D_A \qquad \text{(AGC up-regulation limit constraint)} \tag{4.48}$$

$$\sum_i A_i^- \ge D_A \qquad \text{(AGC down-regulation limit constraint)} \tag{4.49}$$

$$\underline{P_{Gi}} \le P_{Gi} + A_i^+ + R_i \le \bar{P}_{Gi}, i \in I \quad \text{(power output limits of generators)} \tag{4.50}$$

$$0 \le A_i^+ \le 10 v_{Ai} w_i, i \in I \qquad \text{(AGC up-regulation speed constraint)} \tag{4.51}$$

$$0 \le A_i^- \le 10 v_{Ai} w_i, i \in I \quad \text{(AGC down-regulation speed constraint)} \tag{4.52}$$

$$A_i^+ + P_{Gi} \leq \bar{A}_i, \, i \in I \qquad \text{(AGC up-regulation capacity constraint)} \qquad (4.53)$$

$$A_i^- + \underline{A}_i \leq P_{Gi}, \, i \in I \qquad \text{(AGC down-regulation capacity constraint)} \quad (4.54)$$

$$0 \leq R_i \leq \bar{R}_i, \, i \in I \qquad \text{(reserve ramping limit)} \qquad (4.55)$$

$$w_i \in \{0,1\}, \, i \in I$$

A sequential optimization for solving this problem is simple. First, solve the energy dispatch problem; second, solve the AGC dispatch problem; and finally solve the reserve dispatch problem. Obviously, if a solution can be found with this method, the solution is feasible for the co-optimization problem. Now let us look at the energy dispatch model:

$$\operatorname*{Min}_{P_G} \quad \sum_i p_i P_{Gi} \qquad (4.56)$$

$$S.T. \quad \sum_i P_{Gi} = P_D \qquad (4.57)$$

$$\underline{P}_{Gi} \leq P_{Gi} \leq \bar{P}_{Gi}, \, i \in I \qquad (4.58)$$

Similarly, this problem can be easily solved by simply ranking the bid prices. If the transmission constraint cannot be neglected, the model can be solved by a linear programming algorithm. If we assume the optimal solution for energy dispatch is $P_{Gi}^*, \, i \in I$, the AGC dispatch problem is changed into

$$\operatorname*{Min}_{A} \quad \sum_i w_i a_i \left(A_i^+ + A_i^- \right) \qquad (4.59)$$

$$S.T. \quad \sum_i A_i^+ \geq D_A, \quad \sum_i A_i^- \geq D_A \qquad (4.60)$$

$$0 \leq A_i^+ \leq 10 v_{Ai} w_i, \, 0 \leq A_i^- \leq 10 v_{Ai} w_i, \, i \in I \qquad (4.61)$$

$$A_i^+ \leq \bar{A}_i - P_{Gi}^*, \, A_i^- \leq P_{Gi}^* - \underline{A}_i, \, i \in I \qquad (4.62)$$

$$w_i \in \{0,1\}, \, i \in I$$

The above problem can be solved using a mixed integer programming algorithm. Now let A_i^{+*}, $i \in I$, and A_i^{-*}, $i \in I$ denote the optimal solution of the AGC dispatch problem. The reserve dispatch problem then changes into the following linear program and is solved by a standard solver:

$$\underset{R}{Min} \quad \sum_i r_i R_i \tag{4.63}$$

$$S.T. \quad e^T \mathbf{R} - D_R = 0 \tag{4.64}$$

$$\underline{P}_{Gi} - P_{Gi}^* - A_i^{+*} \le R_i \le \bar{P}_{Gi} - P_{Gi}^* - A_i^{+*}, \; i \in I \tag{4.65}$$

$$0 \le R_i \le \bar{R}_i, \; i \in I \tag{4.66}$$

The aim of real-time reserve markets is procuring reserves via a competitive market approach. While the dispatch model of a reserve market is rather straightforward, the pricing of reserves is not and a satisfactory pricing theory has yet to be found.

We end this section by describing another challenge in reserve market design. The fluctuation ranges of electricity loads can vary widely and ideally the charges for different loads should also be different. This has been recognized in Australia's market. However, designing a market with this feature is no easy task. We will return to this point in the next section.

4.4 Compensation without Competition

In Australia's New South Wales electricity market from 1996 through 1999, no ancillary service real-time market existed and ancillary service was procured by contracts. For example, every month or week, a dispatch center determined the generators to provide AGC and reserve services and payments were made to the generators that provided ancillary services.

The essential idea is: (1) the dispatch center dispatches generators to meet the requirements of frequency controls, and (2) payments are made to generators that provide frequency control services. The total frequency control compensation was determined by consultation of the dispatch center and generators and total fees allocated to each of the 8760 hours in a year. This method was employed by the Zhejiang electricity market in China. The virtue of this method is its simplicity and its major drawback is that generators have no incentives to provide AGC service when energy prices in a real-time market are very high.

We now introduce the AGC service compensation method utilized in Australia's New South Wales electricity market. The compensation methods for other services are similar. The calculation process can be divided into three steps:

1. Calculate the AGC payment for the settlement period. Assuming the settlement period is 1 hour, the AGC fee for the settlement period is equal to

$$period\_fee = \frac{12 \text{ million}}{8760} = 1369.9 \tag{4.67}$$

The $12 million represents the total annual AGC fee and may be determined by negotiation.

2. Calculate the unit price for AGC service for the settlement period. Let v_{10} be the ramp limit of generators in 10 minutes, then:

$$AGC\_rate = \frac{interval\_fee}{\sum v_{10}} \tag{4.68}$$

3. Calculate the reward for generators in the settlement period:

$$AGC\_payment = AGC\_rate \; v_{10} \tag{4.69}$$

In addition to the above compensation, a generator also receives lost opportunity cost compensation if it is forced to provide AGC while its energy price is competitive. The drawbacks of the method in the Australia market are as follows:

■ Determining the total AGC service fee can be arbitrary. The relationship between supply and demand is not a factor, in contrast to a market-based approach.
■ The compensation for AGC in some periods tends to be lower when more AGC regulation capacities are committed to cover large load variations. This discourages generators from providing AGC services.

Appendix 4A: Australia National Electricity Market

The ancillary service market in the Australian national market is perhaps the most complex in the world. It implements eight types of frequency control services:

■ Secondary frequency control up- and down-regulation
■ Six-second up- and down-regulation
■ Sixty-second up- and down-regulation
■ Five-minute up- and down-regulation

Despite the complexities, the basic dispatch model for energy and ancillary services is similar to the one described in this chapter. The detailed dispatch model adopted in the Australian national market can be found at http://www.nemmco.com.au/; also see Figure 4.2.

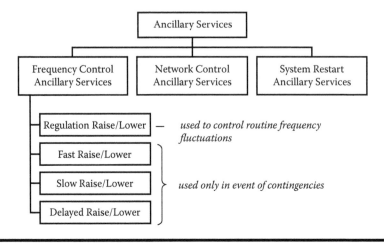

Figure 4.2   Ancillary service segment of Australia national electricity market.

References

D. Gan and E. Litvinov. 2003. Energy and reserve market designs with explicit consideration of opportunity costs. *IEEE Transactions on Power Systems*, 18, 53–59.

N. S. Rau. 1999. Optimal dispatch of a system based on offers and bids: a mixed integer LP formulation. *IEEE Transactions on Power Systems*, 14, 274–279.

Bibliography

W. Cheung, P. Shamsollahi, D. Sun et al. 2000. Energy and ancillary service dispatch for the Interim ISO New England electricity market. *IEEE Transactions on Power Systems*, 15, 968–974.

N. Jaleeli and L. S. VanSlyck. 1999. NERC's new control performance standards. *IEEE Transactions on Power Systems*, 14, 1092–1099.

N. Maruejouls, T. Margotin, M. Trotignon et al. 2000. Measurement of the load frequency control system services: comparison of American and European indicators. *IEEE Transactions on Power Systems*, 15, 1382–1387.

F. C. Schweppe, M. C. Caramanis, R. D. Tabors et al. 1988. *Spot Pricing of Electricity*. Boston: Kluwer.

C. D. Wolfram. 1999. Electricity markets: should the rest of the world adopt the UK reforms? University of California Energy Institute. http://www.ucei.berkeley.edu/PDF/pwp069.pdf

Chapter 5

Market Design: Common Cost Allocations

In an electricity market, it is important for certain common expenses such as transmission and unit start-up costs to be fairly allocated to all market members. To solve these problems, we have to upgrade our mathematical tools. In this chapter, some cooperative game methods are introduced and recommended. Transmission costs and unit start-up costs are introduced and mathematical solutions for allocations are provided. We also study the problem of peaking cost allocation. The concept of transmission rights is also introduced as a method to allocate congestion surplus.

5.1 Background

In an electricity market, many common costs, for example, the imbalance part of income and outlay in market operation, are non-linear functions of electricity market transactions. Generally, these common costs cannot be decoupled. In other words, who is responsible for which segments of common costs is unclear. From the view of a market operator that does not have the objective of earning profits, all common costs must be allocated to market members. Obviously, as deregulation of electric power industry penetrates all over the world, common cost allocation is a major issue that requires the attention of market operators and market members.

As shown in Figure 5.1, when transmission congestion does not occur, the income and outlay of a system are balanced.

The common cost allocation problem is determining who should pay the transmission construction and maintenance costs. Common costs in power systems

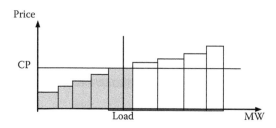

Income = Outlay = CP × Load

Figure 5.1 Clearing of system.

include the imbalance part of income and outlay in market operation, transmission system construction and maintenance costs, voltage support equipment investment costs, ancillary service costs, and others. These costs can be quantified in millions, tens of millions, or hundreds of millions of dollars. Generally, the mapping relationship between common costs and nodal electricity quantity may be analyzed by cost–quantity mapping.

The topology of cost–quantity mapping is very complex and has the characteristics of non-linearity, discontinuity, and large dimension. It can also involve multi-periods. In other words, the information flow covering usage, earnings, and payment of the common costs is uncertain and complicated. A market operator that is not required to earn profits must allocate all common costs to market members in a fair, rational, and complete fashion.

Obviously, the allocation problem involving transmission operation and expansion costs must be solved. However, to simply allocate these costs in proportion to megawatt load is not fair. Common cost allocation continues to be a hot topic in the electricity market area.

The same problem applies to unit start-up cost allocations in pool-based electricity markets. Unit start-up costs incurred to meet system load demand and ensure system security should be allocated completely to market members. However, the relationship between unit start-up and concrete electric loads is not clear. Start-up costs incurred by generators can be allocated proportionally to electrical loads. This approach is easy to implement, has intuitive appeal, and has in fact been used in the industry. However, proportional allocation does not align with the "causer pays" principle. Consider, for example, two loads with identical chronological curves— one in a constrained zone and the other in an unconstrained part of the network. If the start-up costs are allocated in proportion to megawatt load, the load in the unconstrained zone would be subsidizing its counterpart in the constrained zone.

In addition, network losses are produced when all the electricity transactions jointly use a network. Total network loss is the sum of the individual losses (a loss produced by each transaction acts alone) and cross items (produced by transactions acting jointly). A fair, open, and equitable network loss allocation method is essential for the economic interests of market members. The key problem of network loss allocation is how to allocate the cross items. At present, no perfect allocation method for cross items exists.

The unique nature of common cost management lies in the close relationship of economic structure and economic development history in various countries and regional economic bodies. Unique conditions require that each country should develop cost management methodologies and algorithms in accordance with its own economic characteristics and history. In summary, a fair and reasonable allocation method for common cost is essential for the success of an electricity market. Meanwhile, cost allocation is an important component of electricity market theory. In this chapter, we introduce several methodologies for common cost allocation.

5.2 Transmission Costs

Transmission costs fall into two categories: (1) investment in transmission lines and (2) daily operation cost. Obviously, transmission and investment costs should be recovered from network users. For example, in a pool-based electricity market with no transmission congestion, the income from load exactly equals the pay-offs to generators. The income and outlay of the system are balanced. Generally, transmission network companies are strictly regulated. This subsection provides a brief review of various methods of transmission cost allocation and clearly shows the complexity of the problem.

5.2.1 Postage Stamp Methodology

Electric utilities traditionally allocate transmission costs among the users of transmission service based on postage stamp rate methods. With this method, transmission users are not differentiated by the extent of their use of transmission facilities and are charged based on an average embedded cost and the quantity of transacted power. The postage stamp rate is perhaps the simplest method for transmission cost allocation. The calculation steps are as follows:

1. Add the fixed and variable costs for a year.
2. Estimate total wheeled power.
3. Estimate the cost of one unit of wheeled power.
4. Estimate the cost for each wheeling.

The charge is independent of transmission distance when the postage stamp rate method is employed. The transmission cost is determined only by the transacted power. The merits of the method are its simplicity, transparency, ease of implementation, and clarity. The major drawbacks are the failure to reflect correct economic information and inability to ensure economic operation of a power system. However, for a small power system or one with a strong power grid, the postage stamp rate method is a useful choice.

5.2.2 MW Mile Methodology

The MW mile method is a well-known technique for allocating transmission costs. It was proposed by Pacific Gas & Electric (PG&E) in the United States. In the original methodology, DC power flow formulation was used to estimate the use of transmission services. The procedure for multi-transaction assessment is as follows:

1. For transaction t, the transaction related flows on all network lines $MW_{t,k}(t \in K)$ are first calculated using a DC power flow model considering only the nodal power injections involved in that transaction.
2. The magnitude of megawatt flow on every line is multiplied by its length L_k (in miles) and the cost per unit length of the line c_k (in dollars per megawatt mile) and summed over all the network lines as

$$MWMILE_t = \sum_{k \in K} c_k L_k MW_{t,k}$$

The process is repeated for each transaction $t \in T$, including one composed of native generations and loads. Finally, the responsibility of transaction t to the total transmission capacity cost is determined by

$$TC_t = Total\_Cost \frac{MWMILE_t}{\sum_{t \in T} MWMILE_t}$$

The MW mile method considers variations of the power flow in transmission lines and the lengths of the lines. It ensures the full recovery of transmission costs and reflects, to an extent, the actual usage of a transmission system. Therefore, the method is intuitively correct.

5.2.3 Benefit Factors Methodology

The basic idea of benefit factors methodology (Marangon-Lima et al. 1995) is the assigned charge among agents based on the economic "benefit" each one obtains from each network facility. From the perspective of load, the benefit factor of line k to load i is defined as the added electricity fee of load i if line k is out of service. As shown in Figure 5.2, based on benefit factors methodology, the transmission

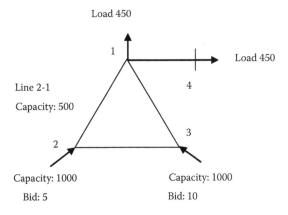

Figure 5.2 Benefit factor methodology example network.

cost of lines 1-2, 1-3, and 2-3 should be charged from load 1 and load 4, while the transmission cost of line 1-4 should be charged from load 4 alone.

5.2.4 Cooperative Game Methodology

The objective of cost allocation (Tsukamoto and Iyoda 1996) is to identify the "causer" market participant (or cost causation) responsible for incurring the costs. In many real-world situations, identifying a causer may be difficult. This is why the cooperative game approach was proposed in the late 1970s. This method does not address the problem of identifying causers directly; rather, it produces a cost allocation that satisfies certain desired properties called axioms. For example, the core allocation satisfies the so-called stand-alone test and the Shapley value allocation satisfies another set of axioms. Therefore, cooperative game methodology has a strong mathematical basis.

As shown in Figure 5.3, three items of a wheeling power contract $N = \{A, B, C\}$ are simultaneously involved in the use of a single transmission line. The load curve of the three items of wheeling power variation over time is shown in Figure 5.4.

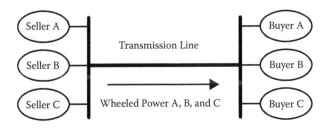

Figure 5.3 Single transmission line with three wheeling transactions.

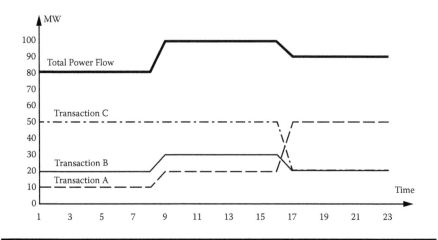

Figure 5.4 Chronological profile of power flow on a transmission line.

When a single wheeler owns the transmission line, the task is to allocate the line cost to wheeling transactions A, B, and C.

We can see from the load curve in Figure 5.4 that each transaction may be divided into three time periods during the 24 hours of a day. Obviously, if Transaction A cooperates with Transaction B, the maximum load flow in each period is 70 MW, which is less than the sum of the individual load flow of Transactions A and B. The load flows of the transmission line under different conditions are

Transaction A uses the transmission line independently: 50 MW.
Transaction B uses the transmission line independently: 50 MW.
Transaction C uses the transmission line independently: 50 MW.
Transaction A cooperates with Transaction B: 70 MW.
Transaction A cooperates with Transaction C: 70 MW.
Transaction B cooperates with Transaction C: 80 MW.
Transactions A and B cooperate with Transaction C: 100 MW.

This analysis shows that when the three transactions cooperate, the maximum load flow of the transmission line reduces from 150 to 100 MW. For ease of illustration, suppose that the transmission fixed cost per megawatt is $120. If each wheeling transaction constructs a transmission line independently, the construction cost would be $C\{A\} = \$6,000$, $C\{B\} = \$6,000$, $C\{C\} = \$6,000$. As shown in Figure 5.4, the peak power flow of each transaction is not superimposed.

Such a cost allocation may be considered a certain type of competitive solution because it involves no cooperation among transactions owners. The economic consequence is too many transmission lines may be constructed. Conversely, if two or three wheeling transactions simultaneously use the same transmission line, the maximum power flow on the line would drop. For example, the maximum energy flow is 80 MW if Transactions B and C form a coalition. As a result, the transmission cost for the two wheeling transactions would be $9,600. The cost function and its value required for a specific subcoalition of these wheeling transactions would be $C\{A,B\} = \$8,400$, $C\{A,C\} = \$8,400$, $C\{B,C\} = \$9,600$. Since the peak in the coalition of wheeling transactions A, B, and C is reduced to 100 MW, the construction cost would be $C\{A,B,C\} = \$12,000$.

Suppose $S = \{A\}, \{B\}, \{C\}, \{A,B\}, \{A,C\}, \{B,C\}, \{A,B,C\}$. We can define the profit function for the coalition as

$$V(S) = \sum_{i \in S} C(i) - C(S)$$

Obviously, for a coalition S, the profit function for the coalition $V(S)$ is equal to its income. For example:

$V\{A\} = \$6,000 - \$6,000 = 0$
$V\{A, B\} = \$6,000 + \$6,000 - \$8,400 = \$3,600$
$V\{A, B, C\} = \$6,000 + \$6.000 + \$6,000 - \$12,000 = \$6,000$

The task is to find z_1, z_2, z_3 that satisfy the following three constraints.

(1) Individual rationality:

$$z_1 \geq V(\{A\}) = 0$$
$$z_2 \geq V(\{B\}) = 0$$
$$z_3 \geq V(\{C\}) = 0$$

(2) Coalition rationality:

$$z_1 + z_2 \geq V(\{A,B\}) = 3600$$
$$z_1 + z_3 \geq V(\{A,C\}) = 2400$$
$$z_2 + z_3 \geq V(\{B,C\}) = 3600$$

(3) Global rationality:

$$z_1 + z_2 + z_3 = V(\{A,B,C\}) = 6000$$

The shaded area in Figure 5.5 indicates the core of this problem where z_1, z_2, z_3 satisfy the constraints. As shown in Figure 5.5, the core is not unique. One approach for solving this problem is utilizing the nucleolus (mass center of the core). Table 5.1

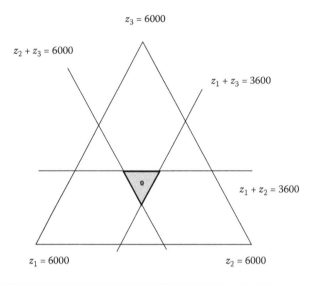

Figure 5.5 Core and nucleolus of transmission cost allocation game.

Table 5.1 Comparison of Cooperative Game and Conventional Methods

	Transaction 1	Transaction 2	Transaction 3
Peak responsibility	2,400	6,000	3,600
Average	12,000/3	12,000/3	12,000/3
Nucleolus	9,600/3	13,200/3	13,200/3

shows the results of employing a nucleolus for transmission cost allocation. A core is a perfect mathematical principle but it has some shortcomings that will be discussed in the next section.

5.3 Unit Start-Up Cost

A valuable characteristic of core allocation is that no individual or subgroup has an incentive to leave a coalition. The center of the core is called the nucleolus, as noted earlier. Cores do not exist in some cases and it is natural to ask under what conditions cores exist. We will discuss the answer in this section. When no core exists, the Shapley value may be employed for common cost allocation. It may be applied in many cases, for example, in analyzing the influences of countries affiliated with the United Nations.

Fixed costs are revealed and recovered using two broad paradigms in electricity markets. Traditionally, energy-only markets are based on the principle of self-commitment and require generators to reflect such fixed costs (along with any capital costs) via energy offer prices. In this paradigm, a generator must offer energy prices and reflect start-up and commitment decisions over the hours by varying the offers. Examples of such energy-only markets include Australia (http://www.nemmco.com.au/), New Zealand, New England (1999 through 2003), Singapore, and East China markets.

Treatment of fixed costs in the electricity market is not perfect in any form of market design. Nevertheless, using energy price alone to enforce commitment decisions is clearly inadequate because generation companies do not know in advance the number of hours their units will be online when they submit bids. As a result, it is difficult for them to average start-up costs into the energy offer price for a period. The result may be inefficient dispatch or, worse, market instability.

The proposed theoretical framework (Hu et al. 2006) for cost allocation relies on an alternative market design in which (1) generators provide a three-part offer composed of energy price, start-up cost, and no-load cost and (2) the independent system operator (ISO) optimizes both commitment and dispatch decisions. A salient feature of this model is that it enables a generation company to better align its offers with its actual cost structure, which includes fixed start-up-related costs and variable energy production costs.

The proponents of three-part bidding opine that it fosters efficient pricing of electricity because it removes distortions driven by energy-only offers. The three-part offer structure also ensures that start-up costs will be fully recovered. This is an important issue that improves the acceptability of this design by market participants. In fact, three-part bidding has been adopted in electricity markets in New England, New York State, and the Northeastern U.S. region. It has also been suggested for inclusion in the standard market design proposed by the U.S. Federal Energy Regulation Council (FERC). Appendix 5B presents some empirical data for the New England ISO market.

Start-up costs incurred by generators can be allocated to electrical loads proportionally. This approach is easy to implement, has intuitive appeal, and has in fact been used in the industry. However, the method does not align with the "causer pays" principle. Consider, for example, two loads with identical chronological curves—one in a constrained zone and the other in an unconstrained part of the network. If the start-up costs are allocated in proportion to megawatt loads, the load in the unconstrained zone would in essence subsidize its counterpart in the constrained zone.

To date little attention has been paid to equitable allocation of start-up costs. Mendes and Kirschen (2000) proposed allocating start-up costs as part of market-clearing prices to avoid allocating the costs directly to loads. Their results are intuitively appealing and have the merit of convenient implementation. Common cost allocation in general has been discussed in the context of electricity markets. While the allocation schemes often rely on engineering principles, a few are based on the axioms of cooperative game theory.

5.3.1 Rationality of Allocations and Cores

To illustrate the problem of cost allocation, a single period case is studied first. Figure 5.6 describes a simple power system comprising three units and three loads. The offer prices of each unit have three components: (1) energy price, (2) start-up cost, (3) and no-load cost.

Table 5.2 lists bid prices of each unit and transmission constraints. Generator G1 bids a low energy price with high start-up and no-load costs. Generator G2 bids a high energy price and low start-up and no-load costs, while G3 bids a high energy price, high start-up costs, and high no-load costs. We assume that demands are fixed (completely price inelastic) so the method has no allocation efficiency implications. While simplistic, this test system is adequate for demonstrating the theoretical concepts. The unit commitment optimization problem for the test problem is as follows:

$$Min \quad \sum_{i=1}^{3} \left(U_i p_i P_{Gi} + U_i \tilde{S}_i + U_i \tilde{N}_i \right) \tag{5.1}$$

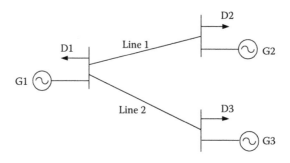

Figure 5.6   Three-unit and three-load power system.

Table 5.2   Bid Prices and Transmission Constraints

Unit/Line	Energy Price ($/MW)	Start-Up Costs ($)	No-Load Costs ($/hour)	Capacity (MW)
G1	12	15,000	500	300
G2	25	12,000	450	250
G3	30	20,000	750	100
Line 1				250
Line 2				100

$$S.T. \quad \sum_{i=1}^{3}(P_{Gi} - D_i) = 0$$

$$\sum_{i=1}^{3} U_i \bar{P}_{Gi} \geq \sum_{i=1}^{3} D_i$$

$$0 \leq P_{Gi} \leq U_i \bar{P}_{Gi}, \ (i = 1, 2, 3)$$

$$F_1(P_{G1}, P_{G2}, P_{G3}) \leq 250 \text{ (line flow constraints)}$$

$$F_2(P_{G1}, P_{G2}, P_{G3}) \leq 100 \text{ (line flow constraints)}$$

$$U_i \in \{0,1\}, \ (i = 1, 2, 3)$$

where $U_i = 1$ indicates that unit i is on, $U_i = 0$ denotes that unit i is off, $\tilde{S}_i$ denotes the start-up cost of unit i, $\tilde{N}_i$ is the no-load cost of unit i, P_{Gi} represents the real power output of unit i, $\bar{P}_{Gi}$ denotes the capability of unit i, D_i is the load at bus i, and p_i is the energy bid price of unit i. All loads and generators are identified by the bus indices.

The line flows F_1 and F_2 are assumed to be loss-less linear functions of generator real power outputs P_{G1}, P_{G2}, and P_{G3}. Demands D_1, D_2, and D_3 are constant parameters in the optimization problem and the minimum output levels of generators are assumed to be zero. Minimum run times and ramp rate constraints of generators are ignored to retain model simplicity. However, incorporating these constraints would not affect the basic conclusions of the study.

Suppose the system loads in a dispatch period are $D_1 = 250$ MW, $D_2 = 200$ MW, and $D_3 = 200$ MW. In this example, load D_3 is located in a constrained area. The optimal dispatch decision is that all three units are called on, incurring a total start-up cost of $47,000, which may be deemed a common cost because it is incurred jointly by all three generators to meet all three loads. The cost allocation problem may be stated as allocating total start-up cost to loads D_1, D_2, and D_3 and ensuring stability, equity, and other desirable results.

Allocation stability means that no subgroup has an incentive to leave the coalition. Equity ensures that no subgroup subsidizes another. The guiding principles of desired results are qualitative and thus not unique. They may even be debatable in relation to practical implementation. For example, a typical approach would be to socialize the cost (allocate it based on proportion) among all loads in much the same way congestion cost is socialized in a pool market (Gan and Chen 2001). However, this may not be acceptable to market participants, as shown by the experience of the New England electricity market. Common cost allocation in other sectors including telecommunications, water rights, and pollution control confronts the fundamental issue that no mutually acceptable solution or best allocation exists. The axioms we discuss in the remainder of this section have desirable properties that improve their acceptability in a market environment.

An alternative to socializing costs is to identify responsibility for each load. This could be measured as the incremental start-up cost a load imposes on a system. One complication in calculating incremental start-up cost is the number of alternative combinations of loads (or coalitions in the parlance of cooperative game theory) that must be considered. Based on the Equation 5.1 model, the incremental start-up costs of all possible coalitions are computed as follows.

If D_3 is the *only* load in the system, G3 must be started and dispatched at full capacity and an additional 100 MW is needed from the other two units. If these 100 MW are provided by G1, the total cost would be $16,700 (12 × 100 + 15,000 + 500). Conversely, if the 100 MW are provided by G2, the total cost would be $14,950 (25 × 100 + 12000 + 450). Hence, the optimal unit commitment and dispatch solution is that G3 and G2 are committed, requiring a total start-up cost of $32,000. Following the same procedure, we find the coalition (C) start-up costs required to meet different combinations of loads:

$C(1) = \$15,000$ (start-up cost of D_1)
$C(2) = \$12,000$
$C(3) = \$20,000 + \$12,000 = \$32,000$
$C(1, 2) = \$15,000 + \$12,000 = \$27,000$ (start-up costs to meet both D_1 and D_2)
$C(1, 3) = \$20,000 + \$15,000 + \$12,000 = \$47,000$
$C(2, 3) = \$20,000 + \$15,000 = \$35,000$
$C(1, 2, 3) = \$20,000 + \$15,000 + \$12,000 = \$47,000$ (start-up costs of all three loads)

To proceed further with this analysis, we must make a transformation. Let L be a possible coalition and $V(L)$ be the associated net benefit or gain of cooperation. Then

$$V(L) = \sum_{i \in L} C(i) - C(L)$$

Using the transformation, we replace a cost allocation problem with a *benefit* allocation problem. Moulin (1988) showed that the two allocation problems are equivalent. We can easily verify that

$V(1) = 0, V(2) = 0, V(3) = 0$
$V(1, 2) = 0, V(1, 3) = 0, V(2, 3) = \$9,000$
$V(1, 2, 3) = \$12,000$

Let X_1, X_2, and X_3 be the gains or benefits allocated to loads 1, 2, and 3, respectively. An acceptable allocation should at least satisfy the following three axioms. First, for a single load, the gain allotted to it should be greater than the gain while it functions alone. This is called the individual rationality axiom: $X_1 \geq V(1)$, $X_2 \geq V(2)$, $X_3 \geq V(3)$. Second, for a coalition, the sum of the gains of its members should be greater than the coalition gain. This is called the coalition rationality axiom: $X_1 + X_2 \geq V(1,2)$, $X_2 + X_3 \geq V(2,3)$, $X_1 + X_3 \geq V(1,3)$. Third, the allocation should satisfy the break-even requirement (all gains must be completely allocated to loads). This is called the global rationality axiom: $X_1 + X_2 + X_3 = V(1,2,3)$.

A set of allocations that meets these three requirements is the core. Usually a core is an area (not a single point). A useful property of core allocation is that no individual or subgroup has an incentive to leave the grand coalition. The center of the core is the nucleolus. As noted earlier, no cores exist in some situations and we must ask under what conditions a core exists. The following theorem (Moulin 1988) explains a sufficient and necessary condition.

Theorem 1—Given a game with N persons, if and only if for all balance coefficients the following inequality holds, the core of the game is not empty:

$$\sum_{S \subset N, S \neq N} \delta_S V(S) \leq V(N) \tag{5.2}$$

The above is also called a balanced game. In Equation 5.2, δ_S is a set of balanced coefficients between 0 and 1 and the following equation is true:

$$\sum_{S:i\in S} \delta_S = 1 \tag{5.3}$$

where δ_S is a value in $\{0, 1, 1/2, 1/3, \ldots, 1/(N-1)\}$. The sum of the balanced coefficients δ_S associated with person i is equal to 1. For example, when $N = 4$, the following inequality is not a condition for the existence of a core:

$$\frac{1}{2}[V(1,2,3)+V(2,3,4)]+V(1,4) \leq V(N)$$

The reason is that the coalitions associated with player 1 are $V(1, 2, 3)$ and $V(1, 4)$ and the sum of the two balanced coefficients is $1/2 + 1 = 3/2 \neq 1$. The same observation holds for player 4. The following inequalities are the ones we must verify:

$$\frac{1}{2}[V(1,2,3)+V(2,3,4)+V(1,4)] \leq V(N)$$

$$V(1) + \frac{1}{2}[V(2,3) + V(2,4) + V(3,4)] \le V(N)$$

$$\frac{1}{3}[V(1,2,3) + V(2,3,4) + V(1,2,4) + V(1,3,4)] \le V(N)$$

The above result leads to the intuitive conclusion that if each of the coalition gain $V(S)$ is relatively small compared to the grand coalition gain $V(N)$, the core of an allocation problem is likely to exist. Based on Theorem 1, we verify the existence of the core of the allocation problem described in this section as follows:

$$V(1) + V(2) + V(3) = 0 \le V(1, 2, 3) = \$12,000$$

$$V(1) + V(2, 3) = \$9,000 \le V(1, 2, 3) = \$12,000$$

$$V(2) + V(1, 3) = 0 \le V(1, 2, 3) = \$12,000$$

$$V(3) + V(1, 2) = 0 \le V(1, 2, 3) = \$12,000$$

$$\tfrac{1}{2}[V(1, 2) + V(2, 3) + V(1, 3)] = \$4,500 \le V(1, 2, 3) = \$12,000$$

Therefore, the core of the allocation problem exists in this case. If a game possesses a unique core, a unique allocation follows directly from it. Unfortunately, this is rarely achieved in most practical problems and there usually exists a continuum of points in the core of a cooperative game. In the next section, we present two methods for finding fair allocation solutions in such situations.

When N is relatively large, the number of inequalities in Equation 5.3 is very large. Super-additivity of a cooperative game can be utilized to reduce the number of inequalities to be verified. Super-additivity is explained as an example. Let S and T be any coalitions of a game and $S \cap T$ be empty. A game is super-additive if $V(S) + V(T) \le V(S \cup T)$. Obviously, if a game is super-additive, we have no need to evaluate the large number of inequalities in Equation 5.2.

It is possible to develop insight on super-additivity by looking at a relatively simple case. The following lemma demonstrates a sufficiency condition under which the start-up cost game defined in this section is super-additive.

Lemma 1—If the capacities and the energy, start-up, and no-load bid prices of all units are identical and the transmission system is unconstrained, the allocation game of start-up cost is super-additive.

Proof of this is included in Appendix 5A. The results indicate that in a system where all units are similar, the cost allocation game is super-additive.

5.3.2 Allocation Based on Nucleolus and Shapley Value

Naturally, the center of the core (or nucleolus) can be viewed as a plausible allocation. Figure 5.7 illustrates the allocation problem discussed in Section 5.3.1. The triangle CDE represents all possible solutions, area ABCD represents the core of the game, and the point *r* at the mass center of the core is the nucleolus. Let *X* be an allocation vector and define the excess function of coalition *S* as

$$e(S, X) = V(S) - \sum_{i \in S} X_i$$

The nucleolus of the game is the solution of

$$\min_{X} \max_{S \subset N} e(S, X)$$

A more detailed description of the nucleolus is available in Moulin (1988).

According to the nucleolus allocation, the coalition gains are obtained as $X_1 = \$1,500$, $X_2 = \$5,250$, and $X_3 = \$5,250$. In other words, loads 1, 2, and 3 are required to pay \$13,500, \$6,750, and \$26,750, respectively. According to the conventional proportional allocation, the three loads would pay \$18,077, \$14,462, and \$14,462, respectively.

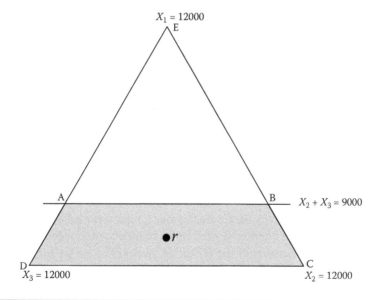

Figure 5.7   Core and nucleolus.

This solution has appeal because it minimizes regret for each player. The value of the excess function is a measure of regret (Tsukamoto and Iyoda 1996). Note that loads 2 and 3 are of the same size and load 3 is located in a tightly constrained area. To meet load 3, the most expensive unit G3 must be committed. To satisfy a coalition without load 3, G3 does not have to be committed. In fact, it can be shown that load 3 should be responsible for start-up costs incurred by G3. The nucleolus allocation clearly reflects this responsibility of load 3.

Conversely, proportional allocation results reflect that load 2 subsidizes load 3. Further, loads 1 and 2 are located in the area without transmission constraints. Thus the larger the load, the higher the allotted cost. The nucleolus allocation correctly reflects this rationality but also has certain shortcomings:

- It may be computationally intensive based on the number of coalitions involved.
- It does not satisfy coalition monotonicity. Each member of a coalition is not assured of increased benefit while the benefit of the grand coalition increases. The incremental benefit of the grand coalition is achieved by contributions of all members and this creates inequity.
- If the allocation game has no core, the nucleolus allocation also does not exist.

The Shapley value has the desirable property of coalition monotonicity. In addition, it satisfies the following axioms (Mas-Colell et al. 1995):

- Efficiency: The break-even requirement of the core is a fundamental requirement.
- Symmetry: the Shapley value depends only on player positions in the game; it does not depend on how players are labeled.
- Linearity: The Shapley value depends linearly on the coefficients $V(S)$ defining the game.
- Zero payoff to dummy players: If a player contributes nothing to the game, its Shapley value is zero.

The formula of the Shapley value allocation is

$$X_i = \sum_s \frac{(|s|-1)!(n-|s|)!}{n!} * [V(s) - V(s - \{i\})] \qquad (5.4)$$

X_i represents the cooperative gain allotted to load i; s denotes a subcoalition including load i; $|s|$ denotes the number of loads in subcoalition s; and n denotes the total number of loads participating in the allocation game. $[V(s) - V(s - \{i\})]$ is equal to the incremental gain of coalition s brought by load i joining the coalition. It follows from Equation 5.4 that each player i is allotted a value equal to its expected marginal contribution across all possible coalitions, which is fair and desirable for all players. Note that $V(0) = 0$ for the numerical example in Section 5.3.1. X_1 can be calculated as follows:

$$X_1 = \frac{0! \cdot 2!}{3!} \left[V(\{1\}) - V(\{1\} - \{1\}) \right]$$

$$+ \frac{1! \cdot 1!}{3!} \left[V(\{1,2\}) - V(\{1,2\} - \{1\}) \right]$$

$$+ \frac{1! \cdot 1!}{3!} \left[V(\{1,3\}) - V(\{1,3\} - \{1\}) \right]$$

$$+ \frac{2! \cdot 0!}{3!} \left[V(\{1,2,3\}) - V(\{1,2,3\} - \{1\}) \right]$$

$$= 1000$$

The rest of the Shapley values are calculated similarly, with X_2 = \$5,500 and X_3 = \$5,500. In other words, loads 1, 2, and 3 will be allotted \$14,000, \$6,500, and \$26,500, respectively. The result of the Shapley value allocation is in the core for this example. However, the example may not always hold. If an allocation does not lie in the core, it is not stable and hence is unattractive. Conditions under which the Shapley value allocation lies in the core are addressed in the Theorem 2 (Moulin 1988):

Theorem 2—If a game is convex, the core of the game is non-empty and the Shapley value allocation is in the core. A game is convex if

$$V(S) + V(T) \leq V(S \cup T) + V(S \cap T), \text{ if } S, \, T \subset N \qquad (5.5)$$

Another definition of a convex game is that if $S \subset T$, S, $T \subset N \backslash \{i\}$, the following inequality holds:

$$V(S \cup \{i\}) - V(S) \leq V(T \cup \{i\}) - V(T) \qquad (5.6)$$

In a convex game, if a load i joins a larger coalition T, it would bring T cooperative gain greater than it would bring to a smaller coalition S. According to Theorem 2, we verify whether the above allocation problem is a convex game as follows:

$$V(1) + V(2) = 0 \leq V(1, 2) = 0$$

$$V(1) + V(3) = 0 \leq V(1 ,3) = 0$$

$$V(2) + V(3) = 0 \leq V(2, 3) = \$9,000$$

$$V(1) + V(2, 3) = \$9,000 \leq V(1, 2, 3) = \$12,000$$

$$V(2) + V(1, 3) = 0 \leq V(1, 2, 3) = \$12,000$$

$$V(3) + V(1, 2) = 0 \le V(1, 2, 3) = \$12,000$$

$$V(1, 2) + V(1, 3) = 0 \le V(1, 2, 3) + V(1) = \$12,000$$

$$V(1, 2) + V(2, 3) = \$9,000 \le V(1, 2, 3) + V(2) = \$12,000$$

$$V(1, 3) + V(2, 3) = \$9,000 \le V(1, 2, 3) + V(3) = \$12,000$$

Therefore, the allocation problem is indeed a convex game. Moulin (1988) proved that a convex game is always a balanced game and this ensures the existence of a core. Lemma 2 demonstrates whether a generalized start-up cost allocation game is convex.

Lemma 2—Let S and T be two load coalitions. Assume no congestion occurs and the capabilities and start-up costs of all generators in the system are the same. Let R_S and R_T be the surplus capabilities that equal the total capabilities of coalitions S and T, subtracting system loads. If for any $S \subset T$, $R_S \le R_T$ always holds, the problem of allocating start-up costs is a convex game.

A proof is included in Appendix 5A. The condition defined in Lemma 2 is very strict because start-up costs is only one of three components in the unit commitment optimization problem. Therefore, when a load joins an existing coalition, we cannot a priori determine the capabilities and start-up costs of generator(s) that need to be committed. Hence, it is generally difficult to ensure conformity to the conditions of a convex game.

The Shapley value has the property of marginality: the marginal contribution of a participant is the only factor that decides its allocation. Thus, the Shapley value is analogous to a marginal cost pricing scheme. At the same time it has desirable properties that make such pricing suitable for equitable and rational fixed cost allocations. Compared to the nucleolus allocation, the Shapley value is somewhat easier to calculate unless the number of coalitions is prohibitively high or an arrangement is unique.

Figure 5.8 compares the allocation results based on the proportional rule, the Shapley value, and the nucleolus method. As discussed in this section, both the Shapley value and the nucleolus approaches give better allocation results than the proportional rule. This demonstrates the usefulness of the suggested axiomatic approaches. Certainly, like any other market designs, the proposed cost allocation principles do not preclude gaming by generators. The motivation behind the principles is to align the generator cost structure with the bidding and further an equitable cost allocation scheme.

We complete this section by noting that the market design described in this section is similar to those currently used in the northeastern U.S. except that this market design is based on an axiomatic approach for allocating fixed costs. The northeastern U.S. markets follow the simple proportional rule. In other words,

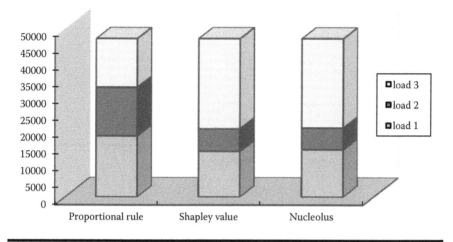

Figure 5.8　Comparison of allocation results.

the energy segment of the market design follows the same dispatch, pricing, and settlement procedures adopted by the U.S. markets.

5.4　Peaking Cost Compensation

Due to the basic requirement for instantaneous balance of supply and consumption, generators should meet base loads and also regulate production to meet load variations. In other words, generators should be capable of regulating production to meet the peaking of electricity supplies.

In a power market environment, the peaking values of generators are sometimes reflected in electricity prices. Alternatively, generators may strategically bid and earn profits based on their peaking capabilities. The Northwest China Power System (NCPS) is rich in hydroelectric resources. Its generation side has been separated from its transmission side and five generation companies were formed. However, the energy market has not yet been established and the electricity prices for generation companies are regulated by the government. The regulated electricity prices for generation companies do not reflect the peaking values of generators. Hence, the peaking values must be quantified, and it is essential to utilize a proper peaking cost compensation mechanism for NCPS generators (Xie et al. 2008). Peaking at NCPS has distinctive characteristics and peaking cost compensation should consider the regulation situation.

In this section, the operations scheduling model of NCPS is formulated. Based on the operations scheduling model, the peaking values of generators are identified. A cooperative game-theory-based peaking cost compensation mechanism for generators is explained.

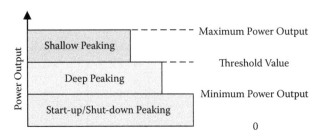

Figure 5.9   Three types of generator peaking.

5.4.1 Peaking Values of Generators

Due to the requirement for simultaneous balance between supply and consumption, generators must meet their base loads and also adjust their production to meet the variations. Generally speaking, as shown in Figure 5.9, generators meet their variational loads via (1) shallow peaking when generators operate at high efficiency, (2) deep peaking when generators operate at low efficiency or in a no-load state, and (3) start-up and shut-down peaking.

The additional peaking cost incurred by hydroelectric units is generally too small to be taken into consideration. However, water discharge losses are incurred when hydro units operate at low power output states, especially during ample inflow seasons. Thermal generators incur no additional cost for shallow peaking although they face some wear-and-tear costs. The peaking costs can be compensated automatically by energy fees. In deep peaking, additional wear-and-tear and fuel costs will be incurred. With start-up and shut-down peaking, more wear-and-tear costs are incurred along with fuel costs. The costs are generally higher because of the needs for heat, compression, and overcoming friction in driving systems.

This section describes the operations scheduling model that demonstrates the scheduling programs at NCPS. The peaking costs of hydro-thermal systems and the peaking values of hydroelectric generators are quantified based on the model.

5.4.1.1 Operations Scheduling of NCPS

Let Tm be the set of indices of scheduling periods (say, $Tm = \{1, 2, \ldots, 24\}$), J be the set of indices of buses, I be the set of indices of thermal units, and H be the set of indices of hydro units. The deep peaking, start-up, and shut-down prices of thermal unit i ($I \in I$) are dp_i, su_i, and sd_i, respectively. The peaking price of water discharge loss of hydro unit h ($h \in H$) is ap_h, and the generation quantity of the unit for the scheduling day is $Q_{H,h}$.

In NCPS, the generation quantities of hydro units are determined by considering the upriver water inflow, the generation quantities of the thermal units, and the regulated electricity prices for hydro and thermal generators. For many reasons, such as social equity issues and adherence to historical contracts, most thermal units are awarded contracts that impose rather tight generation quantity restrictions. Such a contract is represented by $Q_{I,i}$ $(i \in I)$.

Since a large portion of generation quantity of each unit is determined in advance, the operations scheduling of NCPS can be calculated as an optimization problem that minimizes the peaking costs (deep peaking, start-up and shut-down costs of thermal units, and water discharge losses of hydro units) of the whole system:

$$\min \quad \sum_{t \in Tm} \sum_{h \in H} apk_h(t) + \sum_{t \in Tm} \sum_{i \in I} dpk_i(t) + ucu_i(t) + ucd_i(t) \qquad (5.7)$$

where apk_h is the cost due to water discharge loss of hydro unit h; dpk_i, ucu_i, and ucd_i are the deep peaking cost, start-up cost, and shut-down cost of thermal unit i, respectively. The constraints of the optimization problem are shown below.

5.4.1.1.1 Power Balance Constraint

$$\sum_{i \in I} P_{I,i}(t) + \sum_{h \in H} P_{H,h}(t) = \sum_{j \in J} D_j(t), \forall t \in Tm \qquad (5.8)$$

where $P_{I,i}$ is the generation output of thermal unit i. $P_{H,h}$ is the generation output of hydro unit h. D_j is the load demand at bus j.

5.4.1.1.2 Spinning Reserve Constraint

$$\sum_{i \in I} \tilde{P}_{I,i}(t) + \sum_{h \in H} \bar{P}_{H,h}(t) \geq \sum_{j \in J} D_j(t) + R(t), \forall t \in Tm \qquad (5.9)$$

where $\tilde{P}_{I,i}$ is the maximum available power output of thermal unit i. $\bar{P}_{H,h}$ is the maximum power output of hydro unit h. R_j is the spinning reserve demand.

5.4.1.1.3 Transmission Interface Limit

$$-\mathbf{F}_l \leq \mathbf{T}(\mathbf{P}_G(t) - \mathbf{P}_D(t)) \leq \mathbf{F}_l, \forall t \in Tm \qquad (5.10)$$

where matrix $\mathbf{T}$ contains the configuration data and distribution factors of a transmission network. $\mathbf{P}_G(t)$ is the power output of generation units vector at period t. $\mathbf{P}_D(t)$ is the load demand vector at period t. $\mathbf{F}_l$ is the transmission limit vector.

5.4.1.1.4 Feasible Production Region of Hydro Units

$$P_{H,h}(t) \in \pi_{H,h}\left(\underline{P}_{H,h}, \overline{P}_{H,h}, Q_h, \underline{S}_{h,k}, \overline{S}_{h,k}\right), \forall h \in H, \forall t \in Tm \qquad (5.11)$$

where $\pi_{H,h}$ represents the region of feasible production of hydro unit h. $\underline{P}_{H,h}$ is the minimum power output of hydro unit h. $\underline{S}_{h,k}$ and $\overline{S}_{h,k}$ represent the kth upper and lower limits of the feasible power output domain of hydro unit h, respectively.

5.4.1.1.5 Feasible Production Region of Thermal Units

$$P_{I,i}(t) \in \pi_{I,i}\left(\underline{P}_{I,i}, \overline{P}_{I,i}, \tilde{P}_{I,i}(t), Q_{I,i}, RU_{I,i}, UT_{I,i}, DT_{I,i}\right), \forall i \in I, \forall t \in Tm \quad (5.12)$$

where $\pi_{I,i}$ represents the region of feasible production of thermal unit i. $\underline{P}_{I,i}$ is the minimum power output of thermal unit i. $RU_{I,i}$ represents the ramp-up and -down limits of thermal unit i. $UT_{I,i}$ is the minimum up time of thermal unit i. $DT_{I,i}$ is the minimum down time of thermal unit i.

The non-convex, non-linear, mixed integer problem of the operations scheduling model described cannot be solved using conventional non-linear optimization programming. Hence, we require an alternative mixed-integer linear formulation of moderate size that can be solved easily by a standard branch-and-bound method.

5.4.1.2 Quantification of Peaking Values of Generators

As noted above, the four segments of peaking costs of hydro-thermal power systems are deep peaking cost, start-up and shut-down peaking costs of thermal units, and water discharge loss costs of hydro units. The peaking cost PC_1 is defined as

$$PC_1 = \sum_{t \in Tm}\sum_{h \in H} apk_h(t) + \sum_{t \in Tm}\sum_{i \in I} dpk_i(t) + ucu_i(t) + ucd_i(t) \qquad (5.13)$$

When one of the generators of a system operates as a base load unit (evenly distributing the generation quantity among all the scheduling periods), the peaking cost of the system is PC_2. It is obvious that $PC_2 \geq PC_1$, which means that peaking of the generator will lower the peaking cost of the whole system. Hence, the peaking value of the generator is $PC_2 - PC_1$, which is not reflected in the regulated electricity prices for the generator in NCPS, and should be compensated to a certain extent.

When all the generators in a system operate as base load units (evenly distribute the quantities of all the generators among all the scheduling periods), the peaking cost of the system is PC_3. Obviously, the operation scheduling model in Equations 5.7 through 5.12 has no optimum or even feasible solution because the generators all operate as base load units. In this section, PC_3 is determined by solving the following optimization model:

$$PC_3 = \sum_{t \in Tm} \sum_{h \in H} apk_h(t) + \sum_{t \in Tm} \sum_{i \in I} dpk_i(t) + ucu_i(t) + ucd_i(t) + \gamma \sum_{t \in Tm} \sum_{j \in J} L_j(t) \quad (5.14)$$

$$\text{min} \quad PC_3$$

$$\text{S.T.} \quad \begin{array}{c} (5.9)\text{-}(5.12) \\ \sum_{i \in I} P_{I,i}(t) + \sum_{h \in H} P_{H,h}(t) = \sum_{j \in J} D_j(t) + L_j(t), \forall t \in Tm \end{array} \quad (5.15)$$

where γ is the load-shedding price and L_j is the load-shedding amount at bus j. The above optimization model is an extension of the operations scheduling model in Equations 5.7 through 5.12 that may be reformulated as a mixed-integer linear programming problem, and solved using a standard branch-and-bound algorithm.

The load shedding price γ that exerts a critical impact on PC_3 is calculated as follows. In ample inflow seasons, the peaking task is mainly assumed by thermal units. Hence the material cost of peaking in ample inflow seasons reflects the peaking values of generators. We denote the peak load amount of a representative day in ample inflow season as PL and calculate the peaking cost PC of a representative day using Equations 5.14 through 5.15. γ is equal to PC divided by PL:

$$\gamma = PC/PL \quad (5.16)$$

Obviously, $PC_3 > 0$ and $PC_3 > PC_1$, and the peaking value of all the generators is $PC_3 - PC_1$. Since the economical operation of hydro units depends on upriver inflows, their peaking value should be determined as follows. No costs associated with deep, start-up, and shut-down peaking of hydro units will be incurred. Hydro units will be operated at peak load hours, especially in low inflow seasons when their generation quantities are limited. In this way, the peaking cost of the whole system will be reduced and economical operation will be realized. The peaking values of NCPS hydro units vary by season. In low inflow seasons, the peaking values are high. However, in ample inflow seasons, the hydro units should operate as base load units to keep their peaking values low.

5.4.2 Peaking Cost Compensation: Cooperative Game-Based Mechanism

According to the regulatory requirements imposed on NCPS, the task is to design a mechanism based on collecting and reallocating peaking costs to generators. One may argue that the peaking costs of both hydro and thermal generators can be compensated by a cost-based approach. As noted earlier, hydro units do not have material start-up, shut-down, or deep peaking costs. Using a cost-based approach, hydro units will be paid very little, especially in low inflow seasons. Such a solution is not acceptable because the contributions (not the costs) of hydro units for peaking are not reflected properly and a more effective mechanism is needed.

5.4.2.1 Cooperative Game

The model of cooperative game theory has two basic elements: (1) a set of players and (2) a characteristic function. The set of players is composed of all independent benefit entities affecting allocations. If the peaking cost compensation problem involves n ($n > 1$) generators (benefit entities), the set of players (generators) is expressed as $N = \{1,2,\ldots,n\}$. The characteristic function $B(S)$ is a real function of coalitions S, $S \subseteq N$.

The game is to allocate the peaking contribution $PC_3 - PC_1$ among all the generators. The common methods are based on the concepts of core, nucleolus, and Shapley value. In this section, Shapley value is applied to solve the peaking cost compensation problem. Suppose $B(S)$ is the peaking value of coalition S:

$$B(S) = PC(\Phi) - PC(S) \tag{5.17}$$

where $PC(\Phi)$ is the peaking cost when all the generators operate as base load units. Obviously $PC(\Phi) = PC_3$. $PC(S)$ is the peaking cost when coalition S functions alone; obviously $PC(N) = PC_1$. If no feasible solution is found using the operations scheduling model (Equations 5.7 through 5.12) when coalition S functions, the optimization model (Equation 5.15) is used to calculate $PC(S)$. Note that if the operations scheduling model has no feasible solution when S functions, when any coalition $F(F \subset S)$ functions, the operations scheduling model will not have a feasible solution either because more generators operate as base load units. Suppose $V(S)$ is the benefit of coalition S created by the cooperation of its members. Then

$$V(S) = B(S) - \sum_{h \in S} B(h) \tag{5.18}$$

Peaking cost compensation for generators based on the Shapley value can be calculated as

$$\phi_g(V) = \sum_{g \in S} \frac{(n - |S|)!(|S| - 1)!}{n!} \cdot \left[V(S) - V(S - \{g\}) \right], g = 1, 2, \cdots, n, S \subseteq N \tag{5.19}$$

where g represents a generator taking part in peaking cost compensation, $\phi_g(V)$ represents the benefit apportioned to generator g, and

$$\sum B(g) + \phi_g(V) = PC_3 - PC_1$$

S represents the coalitions involving generator g. $|S|$ is the number of units in coalition S. n is the number of units taking part in peaking cost compensation. $n!$ represents the total number of coalition permutations that can be created from the participants in the grand coalition. $S - \{m\}$ represents coalition S without the participation of player m. $V(S) - V(S - \{m\})$ represents the incremental gain that player m accedes to coalition S, namely, $V(S) - V(S - \{m\})$ is the marginal gain of coalition S.

5.4.2.2 Properties of the Peaking Cost Compensation Game

The peaking cost compensation game defined in this section is monotony as Lemma 3 shows—the larger the coalition scale, the higher the coalition benefit. The game is super-additive under certain conditions, as Lemma 3 shows. It is not a convex game, as asserted by Lemma 4.

Lemma 3—The peaking cost compensation game defined in this section is a monotony.

Proof—Let S and R be two generator coalitions, $S \subset R \subseteq N$. The more the generators in the coalition, the lower the peaking cost of the whole system. This implies that $PC(R) \leq PC(S)$. Combining Equation 5.17, the above inequality can be transformed into

$$B(R) \geq B(S) \tag{5.20}$$

which is the desired result. The proof is complete.

Lemma 4—Let S and R be two generator coalitions, $S \subset R \subseteq N$, S and $R \subseteq N$, $S \cap R = \Phi$. If the following inequality holds:

$$V(S \cup R) \leq 0 \tag{5.21}$$

then the net benefit of coalition $S \cup R$ is not more than zero and the problem of peaking cost compensation for generators defined in this section is super-additive:

$$V(S \cup R) \geq V(S) + V(R) \tag{5.22}$$

Proof: The larger the coalition of generators, the lower the peaking cost of the whole system. The peaking cost when coalition $S \cup R$ functions is not more than the peaking cost when coalition S or T functions independently:

$$PC(S \cup R) \leq PC(S), \quad PC(S \cup R) \leq PC(R)$$

Combining with Equation 5.17, the above two inequalities can be transformed into $B(S \cup R) \geq B(S)$ and $B(S \cup R) \geq B(R)$. Summing the above two inequalities, $2B(S \cup R) \geq B(S) + B(R)$. Recalling Equation 5.19, the above inequality can be transformed into $2V(S \cup R) \geq V(S) + V(R)$. Since Equation 5.22 holds, the above inequality means that Equation 5.22 holds and the proof is complete.

Lemma 5—The peaking cost compensation problem defined in this section has the characteristic of decreasing returns to cooperation. Let S and R be two generator coalitions, $S \subset R \subseteq N$ and generator $g \in N/R$. The following inequality holds:

$$V\left(R \cup \{g\}\right) - V(R) \leq V\left(S \cup \{g\}\right) - V(S) \tag{5.23}$$

Proof: The more generators in a coalition, the lighter the peaking task of the whole system. Furthermore, the lighter the peaking task of the system, the lower the peaking value a generator can contribute. In other words, the peaking value of a generator during light peaking is lower than the peaking value of the generator when a peaking task is heavy. Notice that $S \subset R \subseteq N$ and $g \in N/R$. Then

$$PC(R) - PC\left(R \cup \{g\}\right) \leq PC(S) - PC\left(S \cup \{g\}\right)$$

Recalling Equation 5.17, we obtain $B(R) - B(R \cup \{g\}) \geq B(S) - B(S \cup \{g\})$. From the definition of the benefit function in Equation 5.18, the above inequality can be transformed into

$$\sum_{l \in R \cup \{g\}} B(l) + V\left(R \cup \{g\}\right) - \left[\sum_{g \in R} B(g) + V(R)\right]$$

$$\leq \sum_{l \in S \cup \{g\}} B(l) + V\left(S \cup \{g\}\right) - \left[\sum_{l \in S} B(l) + V(S)\right]$$

Rearranging the terms we have Equation 5.23. The proof is complete.

Lemma 6—For a given system, the peaking cost compensation for all the hydro units is greater than that for all the thermal units, i.e.,

$$\sum_{g \in H} x_g > \sum_{g \in T} x_g$$

(x_g is the compensated peaking cost for generator g), if the three conditions below are all satisfied:

1. $PC(H \cup T) = 0$: The peaking cost of the system is zero when all the generators operate as peak load units.
2. $PC(H) = 0$: The peaking cost of the system is zero when all the hydro units
   operate as peak load units and all the thermal units operate as base load units.
3. $PC(T) > 0$: The peaking cost of the system is positive when all the thermal
   units operate as peak load units and all the hydro units operate as base load units.

Proof—For the given system, from *Lemma 3* and $PC(T) > 0$, the following inequality holds:

$$PC(\Phi) \geq PC(T) > 0 \tag{5.24}$$

Let all the hydro units act as an equivalent hydro unit and all the thermal units act
as an equivalent thermal unit. Combining Equations 5.17 through 5.19, we can see
that the peaking cost compensation for the equivalent hydro unit is greater than that
for the equivalent thermal unit $(x_H > x_T)$. In other words, for the given system:

$$\sum_{g \in H} x_g > \sum_{g \in T} x_g \geq 0 \tag{5.25}$$

which is the desired result. The proof is complete.

Lemma 6 implies that by employing the game-based mechanism, the peaking
value of hydro units in low inflow seasons is compensated to an extent. However,
based on a cost-based approach, the units would not be compensated at all.

Lemma 7—If generator g is a base load unit, for coalition S, $S \subset N$, $V(S \cup \{g\}) -$
$V(S) = 0$.

Proof—Because generator g is a base load unit for coalition S and generator g,
$S \subset N, g \in N$.

$$PC(g) = PC(\Phi) \tag{5.26}$$

$$PC(S \cup \{g\}) = PC(S) \tag{5.27}$$

Combined with Equation 5.17, the above two equalities can be transformed into

$$B(g) = PC(\Phi) - PC(g) = 0 \tag{5.28}$$

$$B(S \cup \{g\}) = B(S) \tag{5.29}$$

Recalling Equation 5.18, we can easily obtain the desired result: $V(S \cup \{g\}) - V(S) = 0$. The proof is complete.

Lemma 8—*If the peaking cost of a given system when coalition S (S ⊂ N) functions equal zero [PC(S) = 0], the peaking values of other coalitions contained in S are all equal to that of coalition S. In other words, for coalitions S and R, S ⊂ R ⊆ N, the following equality holds:*

$$B(T) = B(S) \tag{5.30}$$

Proof: Because $PC(S) = 0$, combining *Lemma 3* for coalition T, $S \subset R \subseteq N$, we have $PC(T) = 0$. Recalling Equation 5.17, we obtain the desired result and the proof is complete. With Lemmas 7 and 8 in hand, we can reduce the calculation complexity of the Shapley value.

5.4.2.3 Methods for Reducing Computational Effort in Shapley Value Computation

Applying the Shapley value to cost allocations for bulk hydro-thermal power systems ($n > 10$) imposes an extremely heavy calculation burden. We propose several methods to reduce the calculation burden.

5.4.2.3.1 Simplified Algorithm for Shapley Value

The proposed simplified algorithm for the Shapley value is based on the concept of a bilateral Shapley value. For the gth player (generator), let the other n-1 generator be an equivalent generator, i.e., the peaking cost compensation problem of the n generator is transformed into a peaking cost compensation problem of two generators [$\{g\}$ and $N\backslash\{g\}$]. The benefit apportioned to generator g is easily derived from Equation 5.20 as follows:

$$\phi'_g(V) = \frac{1}{2}\left[V(N) - V(N - \{g\}) + V(\{g\})\right] \tag{5.31}$$

The allocation result determined by Equation 5.31 will probably not ensure zero sum allocation. The imbalance is

$$\delta = V(N) - \sum_{l \in N} \phi'_l(V) \tag{5.32}$$

By allocating the imbalance δ to each generator according to Equation 5.31, a simplified algorithm for the Shapley value is obtained as

$$\phi''_g (V) = \frac{1}{2}\left[V(N) - V(N - \{g\}) + V(\{g\})\right] + \frac{\phi'_g (V)}{\displaystyle\sum_{l\in N} \phi'_l (V)} \cdot \delta \qquad (5.33)$$

The total compensated peaking cost for generator g is

$$x_g = B(g) + \phi''_g (V), g = 1, 2, \cdots, n \qquad (5.34)$$

The simplified algorithm for the Shapley value can reduce the number of coalitions involved in the value calculation significantly. We can use the Northwest China Power System as an example. The system has 268 hydro-thermal units. If the non-simplified algorithm for the Shapley value is employed, the number of coalitions to consider is $C^1_{268} + C^2_{268} + \ldots C^{267}_{268} + C^{268}_{268} = 2^{268} - 1$. However, if the simplified algorithm is employed, the number of coalitions to consider is $C^1_{268} + C^{267}_{268} + C^{268}_{268} = 537$.

5.4.2.3.2 Grouping Related Generators

If each unit belongs to a certain power plant (economic entity), we can group the hydro-thermal units that belong to the same power plant and treat power plants as compensation entities, not as units. Grouping the units belonging to the same power plant (related generators) can greatly reduce the number of coalitions involved in the Shapley value computation. NCPS as an example has 29 power plants in the system. When grouping the related hydro-thermal units according to power plants, the number of coalitions to be considered is $C^1_{29} + C^2_{29} + \ldots C^{28}_{29} + C^{29}_{29} = 2^{29} - 1$. This is in contrast to $2^{268} - 1$, which is significantly larger.

In practice, the two methods proposed for overcoming the combinatorial explosion problem of the Shapley value can be integrated. First, we group the generators belonging to the same power plant and treat the power plants as compensation objects. If more than 10 power plants are involved, the simplified algorithm for the Shapley value is employed. Otherwise, we can use the non-simplified algorithm for the Shapley value.

5.4.3 Peaking Cost Compensation: Engineering-Based Mechanism

The effectiveness of generators in the peaking of the system is determined by the peaking capabilities of generators. In this section, two numerical indices for measuring the peaking capability of generators are devised. The indices are suitable for hydroelectric and thermal units. The proposed peaking cost compensation mechanisms are based on the peaking capability indices. Compared with the axiom-based approach, the advantages of these engineering methods lie in their simplicity and flexibility in application.

5.4.3.1 Indices for Measuring Peaking Capabilities of Generators

The simplest index for measuring the peaking capability of a generator is peaking range, which is determined by the maximum and minimum power outputs of the studied generator as follows:

$$\alpha_g = P_{g\max} - P_{g\min} \tag{5.35}$$

Using the ideal two-generator system in Figure 5.10, suppose the load L is 600 MW, the operating ranges of G1 and G2 are 200 to 600 and 100 to 300 MW, respectively, and the transmission limit is 500 MW. No matter what the other technical parameters of G1 and G2 are, the maximum available power output of G1 is 500 MW, and its minimum must-run power output is 300 MW. It is obvious that because of other system operation constraints, the maximum available power output of G1 is reduced and its minimum must-run power output is increased, thus restricting its peaking capability. Clearly, the peaking capabilities of generators depend on their operating parameters and also on system constraints.

The maximum available peaking range (MAPR) reflects the operation parameters of generators and other system constraints. The MAPR for generator g ($MAPR_g$) is determined by solving the following optimization problem:

$$\tag{5.36}$$

$$\max \quad MAPR_g = \sum_{t=2}^{24} \left| P_g(t) - P_g(t-1) \right|$$

$$S.T \qquad (5.8)\text{–}(5.12)$$

In the above problem, the generation quantity constraint contained in Equation 5.11 is adjusted as

$$\sum_{t \in T} P_{H,g}(t) = Q_{H,g}, \ \forall g \in H \tag{5.37}$$

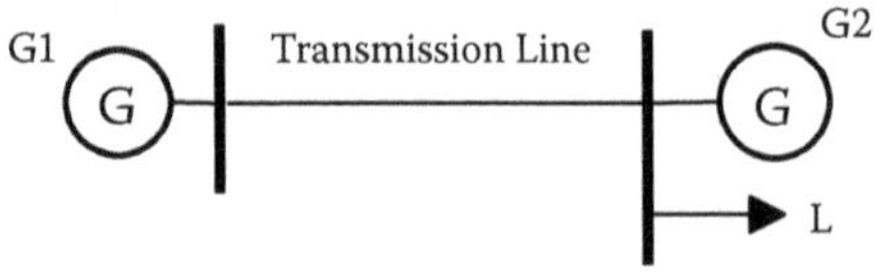

G1   Transmission Line   G2

(G) ————————— (G)

→ L

Figure 5.10   **Single line diagram of two-generator example system.**

The above model is similar to the operations scheduling model in Equations 5.7 through 5.12. The difference lies only in the objective function.

5.4.3.2 Peaking Cost Compensation

The idea is to make the compensated peaking cost for generator g proportional to its peaking capability,

$$x_g = \frac{PKI_g}{\sum_{l \in N} PKI_l}(PC_3 - PC_1), g \in H \cup T \qquad (5.38)$$

where PKI_g is the peaking capability index of generator g. The result could be one of the indices described in this section—peaking range or MAPR. The difference between game-based and engineering-based mechanisms is that they provide different economic signals for generators. In the engineering-based method, the stronger the peaking capability of a hydro generator, the more compensation the generator receives. The compensation results in the game-based method are determined by the marginal contributions of each generator.

5.4.3.3 Simulation Results

NCPS encompasses four provincial power systems: Shanxi, Gansu, Ningxia, and Qinghai. The four provincial systems are interconnected by tie lines as shown in Figure 5.11. The data from a representative day of the low inflow season (winter) in

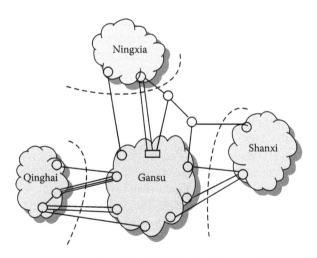

Figure 5.11   Configuration of the Northwest China Power System.

Table 5.3   Ratios of Hydroelectric Unit Capacities

Power System	Shanxi	Gansu	Ningxia	Qinghai
Hydro ratio (%)	10.38	34.95	17.01	78.46

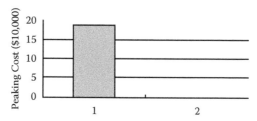

Figure 5.12   Peaking costs of the Northwest China Power System. (1) Base load operation (BaseU). (2) Peak load operation (PeakU).

2007 were used in a simulation study to examine the effectiveness of the proposed method. The load-shedding price for peaking γ was determined using the data from a representative day during the ample inflow season (summer) of 2007. The ratio of hydro generation capacity to total generation capacity of each provincial power system is listed in Table 5.3. Note that Qinghai has a much higher ratio of hydro generation capacity than the others.

Figure 5.12 depicts the peaking costs of NCPS under two scenarios: (1) hydro units operate as base load units (BaseU) and (2) hydro units operate as peak load units (PeakU). The figure indicates that when hydro units operate as peak load units, the peaking cost of the whole system decreases from $183,700 to 0, i.e., due to the peaking of hydro units, no demand is made on the thermal units to start up, shut down, or utilize deep peaking to handle the peaking task. As a result, the peaking value of NCPS hydro units is distinctive during low inflow seasons.

The NCPS has 137 hydro units and 131 thermal units belonging to 10 hydro power plants (HYDPLA) and 19 thermal plants (THEPLA), respectively. Using the 29 power plants of both types (POWPLA) as compensation objects and grouping the related generators, we calculate the peaking capability (MAPR) of generators of each power plant. The peaking cost compensation for each power plant is calculated by the game-based method (Method I) and the engineering-based method (Method II), respectively. The results are shown in Table 5.4. In the engineering-based method, the peaking capability index MAPR is used.

Table 5.4 reveals that the generators that have stronger peaking capabilities receive higher peaking cost compensation. The total compensation for hydro power plants is greater than the total compensation for thermal plants. The peaking value of hydro units is greater than that of thermal units in NCPS.

Unlike the engineering-based method, the game-based method compensates less to generators that have stronger peaking capabilities and more to generators

Table 5.4   Peaking Cost Compensation of Power Plants and Other Simulation Results

POWPLA	MAPR (GW)	Compensation	
		Method I ($)	Method II ($)
HYDPLA-A	999	8005	7700
HYDPLA-B	251	2018	1934
HYDPLA-C	1561	13037	12034
HYDPLA-D	609	4808	4698
HYDPLA-E	401	3289	3088
HYDPLA-F	390	3321	3004
HYDPLA-G	3817	20469	29432
HYDPLA-H	1759	15491	13564
HYDPLA-I	1437	12342	11079
HYDPLA-J	335	2648	2586
THEPLA-A	690	5999	5318
THEPLA-B	959	7994	7391
THEPLA-C	718	6216	5540
THEPLA-D	701	6032	5409
THEPLA-E	475	4160	3664
THEPLA-F	4135	23720	31880
THEPLA-G	233	2123	1797
THEPLA-H	538	4555	4146
THEPLA-I	534	4518	4120
THEPLA-J	719	6031	5540
THEPLA-K	354	3119	2728
THEPLA-L	343	3096	2642
THEPLA-M	179	1569	1377
THEPLA-N	179	1608	1383
THEPLA-O	413	3613	3186
THEPLA-P	332	2891	2559
THEPLA-Q	385	3421	2968
THEPLA-R	227	2024	1749
THEPLA-S	153	1414	1182

with weaker peaking capabilities. In the engineering-based method, the compensated peaking costs for generators are proportional to their peaking capabilities. The compensated peaking costs for generators in the game-based method are determined by their marginal contributions to the coalition. The game-based method yields a sharper economical signal for hydro units that have weaker peaking capabilities.

5.5 Transmission Rights

In the paradigm of nodal price systems, surplus will be produced during market operation. We start this discussion with a two-generator system in Figure 5.13. Based on the figure, if nodal pricing is adopted, the market is not balanced in terms of settlement because (1) sales income is $150 \times 30 + 50 \times 5 = \$4,750$ and (2) payments to generators are $50 \times 30 + 150 \times 5 = \$2,250$.

When transmission congestion occurs, income always exceeds payment. The difference between the income and the payment is called congestion surplus. Congestion surplus is always greater than or equal to zero (Hogan 1992). A proof of this assertion is presented below.

Theorem 3—Congestion surplus is greater then or equal to zero.

*Proof—*Let Ω be the active transmission constraints. Note that

$$congestion\_surplus = \sum_{i=1}^{NB} \rho_i \left(P_{Di} - P_{Gi} \right)$$

$$= \sum_{i=1}^{NB} \left(\lambda + \sum_{k \in \Omega} \mu_k T_{ki} \right) \left(P_{Gi} - P_{Di} \right) \tag{5.39}$$

$$= \lambda \sum_{i=1}^{NB} \left(P_{Gi} - P_{Di} \right) + \sum_{i=1}^{NB} \left(\sum_{k \in \Omega} \mu_k T_{ki} \right) \left(P_{Gi} - P_{Di} \right)$$

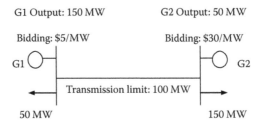

G1 Output: 150 MW   G2 Output: 50 MW

Bidding: $5/MW   Bidding: $30/MW

G1   G2

Transmission limit: 100 MW

50 MW   150 MW

Figure 5.13 **Dispatch and settlement in a two-generator market.**

$$= 0 + \sum_{k \in \Omega} \mu_k \sum_{i=1}^{NB} T_{ki} \left(P_{Gi} - P_{Di} \right)$$

As $\sum_{i=1}^{NB} T_{ki} \left(P_{Gi} - P_{Di} \right) = \bar{F}_k$, we have

$$congestion\_surplus = \sum_{k \in \Omega} \mu_k \sum_{i=1}^{NB} T_{ki} \left(P_{Gi} - P_{Di} \right) = \sum_{k \in \Omega} \mu_k \bar{F}_k > 0 \qquad (5.40)$$

The proof is complete.

We found that the congestion surplus is equal to

$$\sum_{k \in \Omega} \mu_k \bar{F}_k$$

If Ω is empty, the congestion surplus is zero. Allocating market congestion surplus is one of the tasks in market design. The surplus may be allocated in proportion to loads. In North America, the popular method for allocating congestion surplus is the financial transmission right, defined as the right to receive some portion of the congestion surplus.

For example, as shown in Figure 5.14, a market participant in node 2 that owns the financial transmission right may obtain part of the congestion surplus. In other words, it has a transmission right to buy the electric quantity at node 1 or hedge the electricity price risk. The advantage of a financial transmission right is that any participant can obtain the right when it is most needed. Of course, the participant must bid the highest price to buy the transmission right.

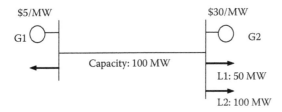

L2 has a 100 MW transmission right, L1 has no transmission right
Nodal price at L2 $5/MW, nodal price at L1 $30/MW

Figure 5.14   Financial transmission right.

A logical way to allocate rights is through simultaneous auction. Now we look at the market model for auctioning. Suppose a market participant wishes to buy the transmission right from nodal i to nodal j with a capacity not greater than $\bar{Z}_l$ MW. The expected price is less than z_l dollars. The objective function of the auction is to meet the needs of all the market participants as much as possible. We let N_Z denote the number of transmission rights and Z_l indicate the winning bids. The objective function of this auction is to maximize

$$\sum_{l=1}^{N_Z} z_l Z_l$$

We find in the two-generator system example that a transmission right should be less than transmission capacity. Otherwise, the market operator will suffer a loss. Transmission right i–j can be simulated as a pair consisting of power input and output. The power inputs at nodal i–j are Z_l and $-Z_l$ accordingly. Obviously, the power flow at the kth branch incurred by this pair of power input and output is equal to $(T_{ki} - T_{kj})Z_l$, as shown in Figure 5.15. As a result, the auction of a transmission right can be formulated as an optimization problem as follows:

$$\underset{Z}{Max} \quad \sum_{l=1}^{N_Z} z_l Z_l \tag{5.41}$$

$$S.T. \quad \sum_{l=1}^{NZ} (T_{ki} - T_{kj})Z_l \le \bar{F}_k, \ (k = 1, 2, \ldots, N_L) \tag{5.42}$$

$$0 \le Z_l \le \bar{Z}_l, \ (l = 1, 2, \ldots, N_Z) \tag{5.43}$$

The above optimization problem can be solved using a standard linear programming (LP) method.

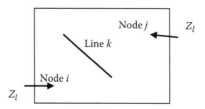

Figure 5.15 Power flow in the kth branch incurred by transmission right i–j.

In an electricity market, the auction of a financial transmission right should be held every month or so. Of course, the coverage of a financial transmission right can also be one year. The transfer of a transmission right is generally node to node in accordance with the characteristics of power systems. In fact, the auction model described is a security-constrained economic dispatch model whose variables are transmission rights. The market participants can buy or sell financial transmission rights through auction or the rights may be traded freely in a secondary market.

In the stated model, we assumed all the transmission rights were sold openly through auction. We indicate a node-to-node transmission right sold at an auction as Z^*. This part of a transmission right is considered constant in the auction model. Then the auction model for transmission rights should be revised as

$$\underset{\mathbf{Z}}{Max} \sum_{l=1}^{N_Z} z_l Z_l \tag{5.44}$$

$$S.T. \quad \sum_{l=1}^{NZ}(T_{ki}-T_{kj})(Z_l+Z_l^*) \le \bar{F}_k, \ (k=1,2,...,N_L) \tag{5.45}$$

$$0 \le Z_l \le \bar{Z}_l, \ (l=1,2,...,N_Z) \tag{5.46}$$

At each trading period in a real-time market operation, market participants that own transmission right Z_l can obtain a payment $(\rho_i - \rho_j)Z_l$. The resulting theoretical problem to be answered is whether the sum of payments is greater than the congestion surplus. The theorem below (Hogan 2002) answers this question.

Theorem 4 (Revenue Adequacy)—In a transmission right auction, congestion surplus is always greater than or equal to zero.

Proof: Note that congestion surplus can be expressed as

$$congestion\_surplus = -\sum_{i=1}^{N_B}\rho_i\left(P_{Gi}-P_{Di}\right)-\sum_{l=1}^{N_Z}(\rho_i-\rho_j)Z_l$$

$$=\sum_{k\in\Omega}\mu_k\bar{F}_k + \sum_{l=1}^{N_Z}\left[\left(\lambda+\sum_{k\in\Omega}\mu_kT_{ki}\right)-\left(\lambda+\sum_{k\in\Omega}\mu_kT_{kj}\right)\right]Z_l$$

$$=\sum_{k\in\Omega}\mu_k\bar{F}_k + \sum_{l=1}^{N_Z}\left[\sum_{k\in\Omega}\mu_k(T_{ki}-T_{kj})\right]Z_l \tag{5.47}$$

$$=\sum_{k\in\Omega}\mu_k\bar{F}_k + \sum_{l\in\Omega}\sum_{k\in\Omega}\mu_k(T_{ki}Z_l-T_{kj}Z_l)$$

$$= \sum_{k \in \Omega} \mu_k \bar{F}_k + \sum_{k \in \Omega} \mu_k \sum_{l \in \Omega} (T_{ki} Z_l - T_{kj} Z_l)$$

$$= \sum_{k \in \Omega} \mu_k \bar{F}_k + \sum_{k \in \Omega} \mu_k F_k$$

$$= \sum_{k \in \Omega} \mu_k (\bar{F}_k + F_k)$$

In the above formula,

$$F_k = \sum_{l \in \Omega} (T_{ki} Z_l - T_{kj} Z_l)$$

is the power flow created by transmission rights. Obviously, the power flow is less than transmission capacity $\bar{F}_k$:

$$\bar{F}_k + F_k \geq 0 \tag{5.48}$$

Due to the fact that $\mu_k > 0$:

$$congestion\_surplus = \sum_{k \in \Omega} \mu_k (\bar{F}_k + F_k) \geq 0 \tag{5.49}$$

The proof is complete.

The two-generator system described previously demonstrates the correctness of the conclusion in the above theorem. Obviously, the congestion surplus is always greater than zero provided that transmission right L2 owns less than or equal to 100 MW. However, if transmission right L2 owns a right greater than 100 MW, the congestion surplus would be less than zero. We now consider the pricing of financial transmission rights. We should use the auction model in vector form. Let $\mathbf{Z} = \{Z_1 \ Z_2 \ \dots \ Z_{NZ}\}^T$ denote the transmission right vector, and

$$\tilde{T}_{kl} = T_{ki} - T_{kj} , \ (l = 1, 2, \dots, N_Z; k = 1, 2, \dots, N_X) \tag{5.50}$$

$\tilde{T}_{kl}$ denotes the sensitivity factor for transmission line k with respect to transmission right l. For all the transmission rights and the transmission lines, the set of such factors can be expressed as $N_L \times N_Z$ matrix $\tilde{\mathbf{T}}$. The auction model can be expressed in matrix form as

$$\underset{\mathbf{Z}}{Max} \quad \mathbf{z}^T \mathbf{Z} \tag{5.51}$$

$$S.T. \quad \tilde{\mathbf{T}} \mathbf{Z} \leq \bar{\mathbf{F}} \tag{5.52}$$

$$0 \leq \mathbf{Z} \leq \bar{\mathbf{Z}} \tag{5.53}$$

Notice that the above auction model has no power balance equations because transmission rights satisfy the power balance condition automatically. The Lagrangian function can be formulated as

$$\Gamma = \mathbf{z}^T \mathbf{Z} + \boldsymbol{\mu}^T [\tilde{\mathbf{T}} \mathbf{Z} - \bar{\mathbf{F}}] + \hat{\boldsymbol{n}}^T (\mathbf{Z} - \bar{\mathbf{Z}}) - \check{\boldsymbol{n}}^T \mathbf{Z} \tag{5.54}$$

The optimum condition is

$$\begin{cases} \dfrac{\partial \Gamma}{\partial \mathbf{Z}} = \mathbf{z} + \tilde{\mathbf{T}}^T \boldsymbol{\mu} - \check{\boldsymbol{\eta}} + \hat{\boldsymbol{\eta}} = 0 \\[2mm] \boldsymbol{\mu}^T [\tilde{\mathbf{T}} \mathbf{Z} - \bar{\mathbf{F}}] = 0, \quad \boldsymbol{\mu} \geq 0 \\[2mm] \hat{\boldsymbol{\eta}}^T (\mathbf{Z} - \bar{\mathbf{Z}}) = 0, \quad \hat{\boldsymbol{\eta}} \geq 0 \\[2mm] \check{\boldsymbol{\eta}}^T \mathbf{Z} = 0, \qquad\quad \check{\boldsymbol{\eta}} \geq 0 \end{cases} \tag{5.55}$$

According to nodal price theory, let $\tilde{\mathbf{T}}^T \boldsymbol{\mu}$ be the transmission right market clearing price vector. Therefore, a transmission right bid is marginally accepted, totally accepted, or rejected and we obtain (1) marginal transmission rights, (2) totally accepted transmission rights, or (3) a rejected transmission right bid, respectively. Obviously, the offer prices of the marginal rights are equal to the market clearing price. The offers of the winning marginal transmission rights exceed the market clearing price and the offers of the rejected transmission rights are less than the market clearing price. The situation is just the opposite of generator bidding.

Now let us consider an example. Suppose a market participant wants to buy the transmission rights in a three-node system as shown in Figure 5.16 as follows:

Transmission right Z_1: nodes 2–1, $z_1 = 99$, $\bar{Z}_1 = 900$
Transmission right Z_2: nodes 3–1, $z_2 = 11$, $\bar{Z}_2 = 100$

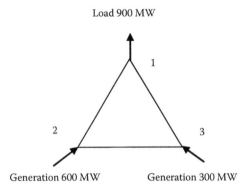

Load 900 MW

1

2

3

Generation 600 MW   Generation 300 MW

Figure 5.16   Three-node system.

Now we set up the auction model. The sensitivity factors of the transmission lines are

$$T = \begin{bmatrix} -0.3333 & 0.3333 & 0 \\ -0.6667 & -0.3333 & 0 \\ 0.3333 & 0.6667 & 0 \end{bmatrix}, \text{ corresponding to transmission lines } \begin{bmatrix} 2-1 \\ 3-1 \\ 2-3 \end{bmatrix}$$

The sensitivity matrix for transmission rights can be formulated as

$$\tilde{T} = \begin{bmatrix} \tilde{T}_{11} & \tilde{T}_{12} \\ \tilde{T}_{21} & \tilde{T}_{22} \\ \tilde{T}_{31} & \tilde{T}_{32} \end{bmatrix}$$

The computational formula for the elements in the matrix is

$$\tilde{T}_{11} = T_{12} - T_{11} = 0.6667; \quad \tilde{T}_{12} = T_{13} - T_{11} = 0.3333$$

$$\tilde{T}_{21} = T_{22} - T_{21} = 0.3333; \quad \tilde{T}_{22} = T_{23} - T_{21} = 0.6667$$

$$\tilde{T}_{31} = T_{32} - T_{31} = 0.3333; \quad \tilde{T}_{32} = T_{33} - T_{31} = -0.3333$$

Therefore, the auction model for a three-node system can be expressed as

$$\underset{Z_1,Z_2}{Max} \quad 99Z_1 + 11Z_2$$

$$S.T. \quad \begin{bmatrix} 0.6667 & 0.3333 \\ 0.3333 & 0.6667 \\ 0.3333 & -0.3333 \end{bmatrix} \begin{bmatrix} Z_1 \\ Z_2 \end{bmatrix} \leq \begin{bmatrix} \bar{F}_{21} \\ \bar{F}_{31} \\ \bar{F}_{23} \end{bmatrix} = \begin{bmatrix} 500 \\ 99999999 \\ 99999999 \end{bmatrix}$$

$$0 \leq Z_1 \leq 900$$

$$0 \leq Z_2 \leq 100$$

We assume that transmission lines 3-1 and 2-3 have very high transmission capacities and therefore the transmission constraints associated with them are not included in the above model. We can verify the solution of the above auction as $z_1 = 750$, $z_2 = 0$, and the price for z_1 is 99.0. In Chapter 3, we calculated the nodal price of a three-node system. The settlement process is

Income from loads: $900 × 15 = $13,500
Payment for generators: $600 × 5 + 300 × 10 = $6,000
Payment for transmission rights: $750 × (15 – 5) = $7,500
Surplus: $13,500 – 6,000 – 7,500 = 0

The result is in accordance with the revenue adequacy theorem.

When some of the transmission rights have been sold in previous auctions, the auction model should be revised. Let $\mathbf{Z}^*$ denote the sold transmission rights. The auction model can be expanded as

$$\underset{\mathbf{Z}}{Max} \quad \mathbf{z}^T\mathbf{Z} \tag{5.56}$$

$$S.T. \quad \tilde{\mathbf{T}}(\mathbf{Z} + \mathbf{Z}^*) \leq \bar{\mathbf{F}} \tag{5.57}$$

$$0 \leq \mathbf{Z} \leq \bar{\mathbf{Z}} \tag{5.58}$$

Appendix 5A: Proofs

Proof of Lemma 3—Let S and T be any two subcoalitions without intersection. Suppose a units are committed when S functions alone, with a coalition gain

$$V(S) = \sum_{i \in S} C(i) - C(S)$$

Let A represent the set of units committed. We have

$$\sum_{i \in S} D_i \leq \sum_{i \in A} P_{Gi}$$

Suppose b units are committed when T functions alone, with a coalition gain

$$V(T) = \sum_{i \in T} C(i) - C(T)$$

Let B represent the set of units committed. We have

$$\sum_{i \in T} D_i \leq \sum_{i \in B} P_{Gi}$$

By hypothesis, if no congestion exists anywhere in the network and all units have the same capacities and bid prices, and the set $A \cap B$ need not be empty, we claim that at most $a + b$ units would be committed when the subcoalition $S \cup T$ functions. In other words, $C(S \cup T) \leq C(S) + C(T)$.

We prove the above assertion by showing that if $a + b + 1$ units are committed, the commitment decision is not optimal for a unit commitment problem (Equation 5.1). Suppose an extra unit j is committed and $j \notin A \cup B$. Let the output of this unit be $P_{Gj} > 0$ (if $P_{Gj} = 0$, it would be uneconomic to commit unit j). The output P_{Gj} is again uneconomic because all the energy prices are identical and should be absorbed by the units in $A \cup B$. The claim is true. Note that

$$\sum_{i \in S \cup T} C(i) = \sum_{i \in S} C(i) + \sum_{i \in T} C(i)$$

Therefore

$$V(S) + V(T)$$

$$= \sum_{i \in S} C(i) - C(S) + \sum_{i \in T} C(i) - C(T)$$

$$= \sum_{i \in S \cup T} C(i) - \left[C(S) + C(T) \right]$$

$$\leq \sum_{i \in S \cup T} C(i) - C(S \cup T) = V(S \cup T)$$

The proof is complete.

Proof of Lemma 4—By hypothesis, the number of generators started up to satisfy coalition T when load i joins is certainly less than when load i joins coalition S. Since start-up costs of all generators are the same, $C(T \cup \{i\}) - C(T) \leq C(S \cup \{i\}) - C(S)$ holds. Based on the definition of benefit function, the above inequality can be transformed into

$$\sum_{j \in T \cup \{i\}} C(j) - V(T \cup \{i\}) - \left[\sum_{j \in T} C(j) - V(T)\right]$$

$$\leq \sum_{j \in S \cup \{i\}} C(j) - V(S \cup \{i\}) - \left[\sum_{j \in S} C(j) - V(S)\right]$$

Since

$$\sum_{j \in T \cup \{i\}} C(j) - \sum_{j \in T} C(j) = C(i), \quad \sum_{j \in S \cup \{i\}} C(j) - \sum_{j \in S} C(j) = C(i)$$

it follows that

$$-V(T \cup \{i\}) + V(T) + C(i) \leq -V(S \cup \{i\}) + V(S) + C(i)$$

That is,

$$V(S \cup \{i\}) - V(S) \leq V(T \cup \{i\}) - V(T)$$

which is the desired result. The proof is complete.

Appendix 5B: Empirical Data for Start-Up Costs

We provide empirical data from the New England ISO electricity market to demonstrate the magnitudes of start-up and no load costs compared to energy costs. Figure 5.17 shows the result using actual data available from the ISO public website (http://www.iso-ne.com). The numbers in the figure represent start-up, no-load, and energy costs September 1–10, 2004 calculated as follows. We assume 50 units were turned on and off every day and the start-up cost for one day was equal to 50 times the average bid start-up costs. Assuming 170 units were on-line each day, the no-load cost for one day was 170 times the average bid no-load costs. Finally, the energy cost for one day was 44 × $16,000 ×24; 44 represents the average price for the month, 16,000 indicates average load, and 24 is the number of hours in a day.

Based on Figure 5.17, the start-up and no-load costs total about 3% of the energy cost. In our example described in Section 5.4, the total energy cost would equal $161,700, while the total start-up and no-load costs would total $62,000. The total start-up and no-load costs thus represent 38% of the total energy cost.

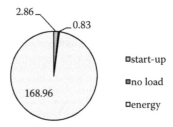

Figure 5.17 Magnitudes of start-up and no-load costs compared with energy costs (in millions of dollars).

This number is different from that for the New England ISO. The reason is that the numbers are hypothetical and were judiciously chosen to highlight the cost allocation principles.

Appendix 5C: Standard Market Design in the U.S.

The four major market features in the celebrated standard market design are (1) locational marginal pricing (LMP), (2) multi-settlement systems, (3) transmission rights, and (4) co-optimization. The time frame and information flow in a standard market design are shown in Figure 5.18.

The tender document for transmission right	⇒	Month-ahead auction for transmission right and contracts	⇒	Allocation of auction revenue
		⇓		
• Load forecasting • Power flow • Offer of unit start up • Offer for active power • Offer for ancillary service	⇒	Day-ahead market auction	⇒	Power output and electricity prices for day-ahead market
		⇓		
• Power flow • Offer for active power • Offer for ancillary service	⇒	Real-time market auction	⇒	Power output and electricity prices for real-time market
		⇓		
Collecton of metered results	⇒	Settlement for real-time market	⇒	

Figure 5.18 Time frame and information flow in standard market design.

References

D. Gan and Q. Chen. 2001. Locational marginal pricing: New England perspective. Presentation summary for panel session on flow gate- and location-based pricing approaches and their impacts. IEEE PES 2001 Winter Meeting, Columbus, OH, January.

W.W. Hogan. 1992. Contract networks for electric power transmission. *Journal of Regulatory Economics*, 4, 211–242.

Z. Hu, L. Chen, D. Gan et al. 2006. Allocation of unit start-up costs using cooperative game theory. *IEEE Transactions on Power Systems*, 21, 653–662.

J.W. Marangon-Lima, M.V F. Pereira, and J. L. R. Pereira. 1995. An integrated framework for cost allocation in a multi-owned transmission system. *IEEE Transactions on Power Systems*, 10, 971–977.

A. Mas-Colell, M. D. Whinston, and J. R. Green. 1995. *Microeconomic Theory*. Oxford, U.K.: Oxford University Press.

D.P. Mendes and D. S. Kirschen. 2000. Assessing pool-based pricing mechanisms in competitive electricity markets. In *Proceedings of Power Engineering Society Summer Meeting*, Vol. 4, pp. 2195–2200.

H. Moulin. 1988. *Axioms of Cooperative Decision Making*. Cambridge, U.K.: Cambridge University Press.

Y. Tsukamoto and I. Iyoda. 1996. Allocation of fixed transmission cost to wheeling transactions by cooperative game theory. *IEEE Transactions on Power Systems*, 11, 620–627.

J. Xie, X. Bai, D. Feng et al. 2008. Peaking cost compensation in Northwest China Power System. *European Transactions on Electrical Power*, 19, 1016–1032.

Bibliography

A. G. Bakirtzis. 2001. Aumann-Shapley transmission congestion pricing. *IEEE Power Engineering Review*, March, 67–69.

R. D. Christie, B. F. Wollenberg, and I. Wangensteen, 2000. Transmission management in the deregulated environment. *Proceedings of IEEE*, 88, 170–195.

J. Contreras and F. F. Wu. 2000. A kernel-oriented algorithm for transmission expansion planning. *IEEE Transactions on Power Systems*, 15, 1434–1440.

J. Contreras and F. F. Wu. 1999. Coalition formation in transmission expansion planning. *IEEE Transactions on Power Systems*, 14, 1144–1150.

W.W. Hogan. 2002. Financial Transmission Rights Formulations. Harvard University, March 31. http://www.ksg.harvard.edu/people/whogan

J. Pan, Y. Teklu, and S. Rahman. 2000. Review of usage-based transmission cost allocation methods under open access. *IEEE Transactions on Power Systems*, 5, 1218–1224.

D. Shirmohammadi, P. R. Gribik, E. T. K. Law et al. 1989. Evaluation of transmission network capacity use for wheeling transactions. IEEE Transactions on Power Systems, 4, 1405–1413.

C.S.K. Yeung, A. S. Y. Poon, and F. F. Wu. 1999. Game theoretical multi-agent modeling of coalition formation for multilateral trades. *IEEE Transactions on Power Systems*, 14, 929–935.

Chapter 6

Microeconomic Analysis

One may wonder what the outcome may be after an electricity market "goes live"—for example, the degree of market power, the trend of electricity prices, and other factors. In this chapter, we first use a microeconomic method known as game theory to analyze these factors. Game theoretic models for studying electricity market behaviors are built to analyze equilibrium and market power. We also analyze the results of two auction experiments in which graduate students participated. We can see the small difference between theoretical analysis and experiment results. This indicates that building an experimental system to analyze complex electricity market problems can be meaningful.

6.1 Background

In the previous chapter, we discussed the common cost allocation problems in electricity markets and found that it is not easy to allocate common costs in electricity markets reasonably. In fact efficiently allocating social resources is a difficult task. To date, engineers and mathematicians have been unable to develop resource allocation methods that are universally applicable. At present, using market mechanisms to allocate social resources is the mainstream method.

Earlier chapters focused on fundamental electricity market structures. Naturally, we wonder what the operational performance of an electricity market would look like? The question involves many factors such as market power, electricity prices, market stability, and market risks. Many methods can be utilized to analyze these matters and we will discuss several of them in this chapter.

Not all markets achieve success. The symbols of success in an electricity market include, for example, decreases in electricity prices, improvement in customer service, and wide application of new technologies and innovations. In our view, the success of an electricity market depends on both physical and human factors. Examples of physical factors are advantages based on economies of scale, market entry and exit barriers, geographic distribution of product, and price elasticity of loads.

The human factors are basically the market rules. The California and New England electricity markets in the United States have made major revisions to their market rules more than once. Minor revisions have never stopped. Continued analysis and adjustment of market rules ensures that they continue to be effective.

Microeconomics can be useful for research concerning electricity market rules. In microeconomics, the concept of strategic gaming relates closely to electricity market theory. Therefore, in this chapter we will discuss the fundamentals of game theory and then introduce economic analysis models for electricity markets.

One common method for microeconomic research is empirical analysis. This technique employs statistics and econometrics to analyze market behaviors (for example, fluctuating electric prices). This method is not covered in this book but its importance is unquestionable. From empirical analysis we can obtain many meaningful results. Our interest in this chapter, however, is to demonstrate the usefulness of game theoretic approaches.

6.2 Fundamentals of Non-Cooperative Game Theory

In Chapter 2, we explained economic dispatch and unit commitment issues and briefly introduced the fundamentals of optimization. A mathematical optimization problem can be described by two factors: (1) an objective function and (2) equality and inequality constraints. An optimal solution can be found in the feasible solution set that makes the objective function optimal. This problem can be concisely expressed as $\max_{s \in S} u(s)$, where S is the feasible set often represented by a group of equality or inequality constraints and $u(s)$ is the objective function.

This example represents a simple decision making scenario, while a game constitutes a more complex problem. Fundamentally, a game involves multiple optimization problems. Games profoundly affect our daily lives at many levels, for example, an arms race, a chess game, or a market negotiation. Games are classified as normal forms or generalized forms. In this book, we focus on the normal form game. Basically, a *normal form game has n decision makers*. Each decision maker faces an optimization problem. Let R denote a conventional Euclidean space. For decision maker i, the feasible set of the optimization problem is S_i (sometimes called the strategy set). The objective function is

$$u_i : S_1 \times S_2 \times ... \times S_n \to R \tag{6.1}$$

The objective function is often called a utility function. A game G can be described as

$$G = \{S_1, S_2, ..., S_n, u_1, u_2, ..., u_n\} \tag{6.2}$$

Or, more concisely:

$$G = \{\mathbf{S}, \mathbf{u}\} \tag{6.3}$$

A game starts when all decision makers choose their respective decisions $s_k \in S_K$; then each decision maker obtains a utility $u_i(s_1, s_2, ..., s_n)$. This is what we call simultaneous game (chess is a sequential game). Let us consider a three-person game problem described by Aliprantis and Chakrabarti (2000). The feasible solution set of the decision makers is the whole real number axis. The utility function is defined as

$$u_1 = 2xz - x^2 y$$

$$u_2 = \sqrt{12(x+y+z)} - y$$

$$u_3 = 2z - xyz^2$$

Our main interest in a game is the behavior of each decision maker. At present, the most popular solution method is called Nash equilibrium. When applied to a normal form game, the Nash equilibrium is a decision solution that satisfies the following set of inequality constraints:

$$u_i(s_1^*, s_2^*, ..., s_{i-1}^*, s_i^*, s_{i+1}^*, ..., s_n^*) \geq u_i(s_1^*, s_2^*, ..., s_{i-1}^*, s_i, s_{i+1}^*, ..., s_n^*) \tag{6.4}$$

$$s_i \in S_i, \ i = 1, 2, ..., n$$

A Nash equilibrium has the following property: for each decision maker, if the rivals adopt Nash equilibrium strategies, the decision maker should also employ the Nash equilibrium strategy. In other words, a Nash equilibrium is a combination of the optimal strategies of all the participants. This means that a participant has no incentive to select another strategy if the other decision makers' strategies are decided. In summary, no participant has an incentive to choose a decision outside the equilibrium.

The Nash equilibrium can be explained from another view. Suppose all the participants in a game reach an agreement that specifies the behavior norms of each participant and there is no force outside the game. Will the participants obey the agreement? In other words, is the agreement self-enforcing? If the participants obey the agreement in a self-enforcing way, the agreement forms a Nash equilibrium: if all participants obey the agreement, no participants have incentives to violate the agreement.

Now, let us analyze the example. Obviously, for decision maker 1, if decision maker 2 and decision maker 3 both employ Nash strategy y^* and z^*, decision maker 1 should solve the optimization problem $\max\limits_{x} u_1 = 2xz^* - x^2 y^*$, or:

$$\frac{\partial u_1}{\partial x} = 2z^* - 2xy^* = 0 \tag{6.5}$$

Decision maker 2 and decision maker 3 face similar optimization problems, respectively,

$$\max_{y} \quad u_2 = \sqrt{12(x^* + y + z^*)} - y \tag{6.6}$$

$$\max_{z} \quad u_3 = 2z - x^* y^* z^2 \tag{6.7}$$

To find a Nash solution of the above game, we solve the following non-linear simultaneous equations:

$$\begin{cases} \dfrac{\partial u_1}{\partial x} = 2z - 2xy = 0 \\[2mm] \dfrac{\partial u_2}{\partial y} = \sqrt{\dfrac{3}{x+y+z}} - 1 = 0 \\[2mm] \dfrac{\partial u_3}{\partial z} = 2 - 2xyz = 0 \end{cases} \tag{6.8}$$

In economics, an important game model is the Cournot model (Jehle and Reny 2000). Under certain conditions, this model can prove that the more suppliers in a market, the higher the market price. As the number of suppliers tends to infinity, the market price moves toward the marginal cost. The celebrated model was proposed by French economist Auguste Cournot in 1838.

Suppose a market has J suppliers and each supplier has a constant unit production cost, i.e., the cost function of supplier j is

$$C(q_j) = cq_j, c \geq 0, \ j = 1, 2, \cdots, J \tag{6.9}$$

where c is the unit production cost and q_j is the output of supplier j. The inverse demand function is expressed as

$$p = a - b \sum_{j=1}^{J} q_j, \quad a > 0, \ b > 0, \ a > c \tag{6.10}$$

where p is the market price and $\sum_{j=1}^{J} q_j$ is the market demand. Therefore the profit of supplier j is

$$\Pi_j(q_1, q_2, \cdots, q_J) = (p - c)q_j$$
$$= (a - b \sum_{k=1}^{J} q_k)q_j - cq_j \tag{6.11}$$

How do we find the Nash equilibrium $(\bar{q}_1, \bar{q}_2, \ldots, \bar{q}_J)$? Obviously, if $q_k = \bar{q}_k$, $k \neq j$; for the sake of profit maximization, supplier j will choose its output $\bar{q}_j$ that satisfies:

$$a - 2b\bar{q}_j - b \sum_{k \neq j} \bar{q}_k - c = 0 \tag{6.12}$$

We then rewrite the above formula:

$$\bar{q}_j = a - c - b \sum_{k=1}^{J} \bar{q}_k \tag{6.13}$$

Notice that the right sides of the above equations are identical for all the suppliers and thus the outputs of all suppliers are equal. Let $\bar{q}$ be the output of an individual supplier. From Equation 6.13, we obtain:

$$\bar{q} = \frac{a - c}{b(J + 1)} \tag{6.14}$$

As a result, the following formulas are satisfied for market equilibrium:

$$\bar{q}_j = \frac{a-c}{b(J+1)} \quad j=1,2,\cdots, J$$

$$\sum_{j=1}^{J} \bar{q}_j = \frac{J(a-c)}{b(J+1)}$$

$$\bar{p} = a - \frac{J(a-c)}{J+1} < a \tag{6.15}$$

$$\bar{\Pi}^j = \frac{(a-c)^2}{b(J+1)^2}$$

The difference between the market price and the marginal cost at the equilibrium point is

$$\bar{p} - c = \frac{a-c}{J+1} > 0 \tag{6.16}$$

Obviously, when $J = 1$ (under the condition of absolute monopoly), the market price is greater than the marginal cost. As $J \to \infty$, we easily find that

$$\lim_{J \to \infty} (\bar{p} - c) = \lim_{J \to \infty} \frac{a-c}{J+1} = 0 \tag{6.17}$$

Based on the above formula, as the number of suppliers in a market approaches infinity, the market price approaches the marginal cost. This result is in accordance with intuition. If a market has two suppliers, $b = 1$, the optimal output of the two suppliers is

$$\bar{q}_1 = \bar{q}_2 = \frac{1}{3}(a-c) \tag{6.18}$$

The profit is

$$\bar{\Pi}^1 = \bar{\Pi}^2 = \frac{(a-c)^2}{9} \tag{6.19}$$

However, when the market is a monopoly, the optimal production is

$$\bar{q} = \frac{1}{2}(a-c) \tag{6.20}$$

The profit is

$$\overline{\Pi} = \frac{(a-c)^2}{4} \tag{6.21}$$

Obviously, the total production of oligopoly competition is greater than that of monopoly competition while the total profit of oligopoly competition is less than that of monopoly competition. The reason is that when each supplier selects its optimal production, it considers only the impact of its production on its own profit and ignores the effects on other suppliers.

Another famous economics game model is the Bertrand model. Under certain conditions it can prove that when the number of suppliers in a market is greater than or equal to 2, the market price equals the marginal price. The model proposed by Bertrand in 1883 is a type of price competition model. A Cournot model is a quantity competition model.

Suppose the demand is $D(p)$ and price competition exists between two suppliers in a market. The production of the two suppliers is perfectly substitutable and the unit production costs of both suppliers are identical and constant. The suppliers simultaneously decide their price p_1, p_2. The production of supplier j is

$$q_j(p_j, p_k) = \begin{cases} D(p_j) & (\text{if } p_j < p_k) \\ \dfrac{1}{2} D(p_j) & (\text{if } p_j = p_k) \\ 0 & (\text{if } p_j > p_k) \end{cases} \tag{6.22}$$

Therefore, when the price of supplier j is p_j, the profit of supplier j will be $(p_j - c)$ $q_j(p_j, p_k)$. Does a Nash equilibrium of the market exist? If a Nash equilibrium exists, is the equilibrium point unique? The following classic result completely answers these questions.

Proposition: in the Bertrand two-player game, a Nash equilibrium exists and the equilibrium point $\overline{p}_1 = \overline{p}_2 = c$ is unique.

Proof: First let us prove that $p_1 = p_2 = c$ is a Nash equilibrium. At this price, both suppliers' profits are zero. If the price of a supplier remains unchanged, can the other supplier increase its profit by changing its price?

1. If the price is greater than c, the production of the supplier is zero and correspondingly the profit is also zero.
2. If the price is less than c, the supplier will produce at a loss. Therefore, $p_1 = p_2 = c$ is a Nash equilibrium.

Second, let us prove that the Nash equilibrium point is unique.

1. Suppose a Nash equilibrium satisfies $\min(p_j,p_k) < c$. If supplier j's price $p_j \leq p_k$ or $p_j < c$, the supplier will produce at a loss. If the price $p_j = c$, its profit will be zero. Obviously, the supplier has motivation to change (in fact increase) its price. Therefore, it is impossible that a Nash equilibrium exists and satisfies $\min(p_j,p_k) < c$.

2. Suppose a Nash equilibrium satisfies $p_j = c$, but $p_k > c$. The profits of suppliers k and j are both zero. However, if supplier j increases its price a little to p'_j, $p'_j = c + \varepsilon < p_k$, for example, its profit will be greater than zero and the Nash equilibrium does not exist.

3. Suppose a Nash equilibrium satisfies $p_j > c$, $p_k > c$. Without loss of generality, let us suppose $p_k \geq p_j$. The maximal profit of supplier k is

$$\frac{1}{2}(p_j - c)D(p_j)$$

However, if the supplier decreases its price, for example, to $p_k = p_j - \varepsilon > c$, its profit will be

$$(p_j - \varepsilon - c)D(p_j - \varepsilon) > \frac{1}{2}(p_j - c)D(p_j)$$

Therefore, the Nash equilibrium does not exist.

As a result, a Nash equilibrium exists in the above two-player Bertrand game and the equilibrium point is unique.

The Bertrand model affirms that, under certain conditions, when the number of suppliers in a market is greater than or equal to two, the market price is equal to the marginal price. The methods described in this section are from Aliprantis and Chakrabarti (2000) and Mas-Colell et al. (1995).

6.3 Game Models for Market Analysis

Predicting and analyzing the performance of electricity market design is an essential subject for the success of an electric power market. The accuracy of analyses and predictions impacts many important aspects of electricity market decision making. Because ex post electricity market analysis may be very costly, many economists and politicians prefer the mainstream game theoretic method that allows prediction of market performance because the players in a game will adopt Nash equilibrium strategies. Some commonly studied methods include:

- Supply function equilibrium models (Green and Newbury 1992)
- Cournot models (Cardell et al. 1997, Wen and David 2001]
- Extended Bertrand models (Gan and Bourcier 2002)

The micro-structure of an electricity market is complex. To improve the prediction abilities of game models, it is preferable to adopt analytical models other than computational types. However, an analytical model that is overly complex will not provide practical insights. To build a proper game model that is analytically manageable, we must make a trade-off between modeling ability and prediction ability. A supply function equilibrium model has found application in English electricity markets. The Cournot model is a traditional economic analysis method suitable for analyzing the long-term performance of electricity markets. However, the Bertrand model is considered more suitable for analyzing short-term performance.

Based on the above considerations, we build a price competition model for oligopolic competition in electricity market auctions. The model can be called the extended Bertrand model. As we know, capacity constraint is an important factor that affects the bidding behaviors of market participants. A key characteristic of the game model described below is its flexibility in modeling capacity constraints.

We first introduce the market dispatch auction model, then discuss an extended Bertrand game model. Finally, the results of equilibrium analysis are introduced. In the game models introduced in this section, all the generators' marginal costs are assumed to be identical.

6.3.1 System Dispatch Model

A power dispatch and pricing solution can be obtained from the following optimization problem:

$$\underset{P}{Min} \quad p^T P \tag{6.23}$$

$$S.T. \quad e^T P = D_P \tag{6.24}$$

$$0 \le P \le \bar{P} \tag{6.25}$$

where $\bar{\mathbf{P}}$ is the high operating limit vector, $\mathbf{e}$ is a vector with all ones, and D_p is the total load. The Lagrangian function of the problem is

$$\Gamma = p^T P + \rho(D_P - e^T P) - \breve{\tau} P + \hat{\tau}(P - \bar{P}) \tag{6.26}$$

where, $\rho, \breve{\tau}, \hat{\tau}$ represent Lagrangian multipliers. These multipliers and the dispatch P must satisfy the so-called Kuhn–Tucker optimality conditions. We know the bid price of the last accepted generator is the market clearing price that equals ρ.

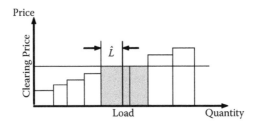

Figure 6.1 Residual load when two generators bid the same price. CP = clearing price. $\hat{L}$ = residual load, i.e., the total load minus the non-marginal generation.

If an auction problem has multiple optimal solutions, it implies that the bid prices of some units are the same and these generators set the clearing price. Under such circumstances, these generators are dispatched in proportion to their capabilities. This widely accepted method is important and therefore is presented mathematically (for illustration in the following equation we assume that the bid prices of two generators are the same):

$$P_1 = \frac{\overline{P}_1}{\overline{P}_1 + \overline{P}_2} \hat{L} \tag{6.27}$$

$$P_2 = \frac{\overline{P}_2}{\overline{P}_1 + \overline{P}_2} \hat{L} \tag{6.28}$$

where $\hat{L}$ is the residual load that the two generators supply (Figure 6.1). The concept is also applicable to a power system with multiple generators.

6.3.2 Game Model

The following assumptions are made in this section.

1. We assume that $e^T \overline{P} - D_p \geq 0$. This ensures feasible solutions to the auction problem.
2. Each unit has a constant marginal cost that is publicly known. This assumption is only an estimate in the real world where generator types and fuel costs are public information.
3. All generators are subject to a price cap $\overline{p}$, $p \leq \overline{p}$.
4. The price cap exceeds production cost $\overline{p} > c$.
5. A generator is allowed to bid one quantity block only. This assumption is not critical to the results; it is needed only for ease of explanation.
6. The game has at least two suppliers, each of which owns generators with non-zero capability.

The payoff function of the game is assumed to be the profit of the supplier. The problem faced by the kth player is choosing a price vector p_k to maximize its profit, assuming that the price vectors of the other suppliers p_{-k}^* are given. If $\mathbf{c}_k$ represents the cost of generators of the kth supplier, the profit of the kth supplier is

$$\pi_k(p_k) = (\rho - c)^T P_k \tag{6.29}$$

Notice that P and ρ are single-valued implicit functions of p_k. Let us denote these functions as $P(\cdot)$ and $\rho(\cdot)$, respectively. Now the profit maximization problem can be reformulated as

$$\underset{p_k}{Max} \quad \pi_k(p_k) = [\rho(p_k) - c]^T P_k(p_k) \tag{6.30}$$

$$S.T. \quad c \le p_k \le \bar{p}_k \tag{6.31}$$

Obviously, the kth supplier's best response p_k is generally a multi-valued function (correspondence or point-to-set map) of its rival's strategies, p_{-k}^*. Let $r_k(\cdot)$ denote this multi-valued function. It follows that

$$\mathbf{p}_k^* \in r_k(\mathbf{p}_{-k}^*)$$

The intersection(s) of the graph of these correspondences $\mathbf{p}_k^* \in r_k(\mathbf{p}_{-k}^*)$ constitute Nash equilibrium. If we rewrite $\mathbf{p}_k^* \in r_k(\mathbf{p}_{-k}^*)$ as a matrix $\mathbf{p} \in \mathbf{r}(\mathbf{p})$, the intersection(s) of the multi-valued function $r_k(\cdot)$ ($k = 1, 2, \ldots, N_{suppliers}$) yield the Nash equilibrium of the game. In other words, the reaction correspondence or function $\mathbf{r}(\cdot)$ has a fixed point. Based on the terminology of game theory, we call $\pi_k(\cdot)$ a profit function and $r_k(\cdot)$ a reaction correspondence.

6.3.3 Equilibrium Analysis

We now introduce equilibrium analysis. When the capacity constraint is very tight, the clearing price at equilibrium is likely to be high because essentially all suppliers are needed and large suppliers therefore can raise the clearing price. Figure 6.2 illustrates such a situation where a large supplier that owns the shaded generators has an incentive to offer its capacity at a price equal to the price cap.

   Lemma 1 (existence of critical load demand)—There exists a critical load level D_P^* such that if $D_P > D_P^*$, the auction game possesses an equilibrium at which $\rho^* = \bar{c}$ (Gan and Shen 2004).

   Proof—Because the objective is to prove the existence of D_P^*, we can hypothesize that if we randomly add bid blocks of generators, except for one generator that is partly dispatched, the other generators will be accepted wholly. Suppose the

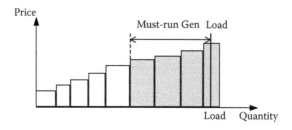

Figure 6.2 Competition under tight capacity constraints.

partly selected generator is numbered m and consider the condition that the mth generator bids cap price while the other suppliers bid marginal cost. Obviously, generator i ($i \neq m$) has no motivation to bid a price that deviates from marginal cost. Under that condition, the profit of the mth generator can be calculated as

$$\pi_m(D_P) = (\bar{p} - c)\left(D_P - \sum_{j \neq m} \bar{P}_j\right) \tag{6.32}$$

We can then find D_P^*, enabling $\pi_m(D_P^*) > 0$. If the bid price of the mth generator is between c and $\bar{p}$, the profit obtained would be less than $\pi_m(D_P^*)$. If the bid price of the generator equals c, all the generators will be dispatched in proportion to their capacities. The market clearing price will be c, and the profit of the mth generator will be zero. We see that the condition established by the mth generated bid is in equilibrium and $\rho^* = \bar{p}$. The proof is complete.

Consider a power system with three gamers and three generators. Suppose $\bar{P}_1 = 100$, $\bar{P}_2 = 200$, $\bar{P}_3 = 500$, and $D_P^* = 300$. The above conclusion is qualitative. Is there a sharper result? The following proposition gives a more concrete characterization of equilibrium points of the game. Before stating the proposition, a useful result should be proved.

Lemma 2 (price competition)—Consider a power dispatch configuration (P^*, ρ^*). If the clearing price $\rho^*(\rho^* > c)$ is determined by a generator owned by several players and the total load is lower than the total generation capacity, this configuration is not at equilibrium (Gan and Shen 2004).

Proof—Without loss of generality, consider the condition in which the clearing price is determined by two generators (generator 1 and generator 2). Generator 1 is owned by player C and generator 2 is owned by player D. $\hat{L}$ is the residual load. The profit of supplier C is

$$\dot{\pi} = \sum_{j \in C, p_j < \rho^*} (\rho^* - c)\bar{P}_j + (\rho^* - c)\hat{L}\frac{\bar{P}_1}{\bar{P}_1 + \bar{P}_2} \tag{6.33}$$

The first term is the profit earned by player C's inframarginal generator and may be zero. Notice that $\hat{L}$ can be greater or less than $\bar{P}_1$. Now, suppose the bid price of generator 1 decreases to $\rho^* - \eta$ where η is a small positive number. First, consider $\hat{L} \leq \bar{P}_1$. The new clearing price (*CP*) is changed to $\rho^* - \eta$ and the profit of gamer C is

$$\ddot{\pi} = \sum_{j \in C, p_j < \rho^*} (\rho^* - \eta - c)\bar{P}_j + (\rho^* - \eta - c)\hat{L}\frac{\bar{P}_1}{\bar{P}_1 + \bar{P}_2} \tag{6.34}$$

The difference between $\dot{\pi}$ and $\ddot{\pi}$ is

$$\ddot{\pi} - \dot{\pi} = -\eta\left(\hat{L} + \sum_{j \in C, p_j < \rho^*} \bar{P}_j\right) + (\rho^* - c)\hat{L}\frac{\bar{P}_1}{\bar{P}_1 + \bar{P}_2} \tag{6.35}$$

Because

$$\left(\hat{L} + \sum_{j \in C, p_j < \rho^*} \bar{P}_j\right) > 0$$

and

$$(\rho^* - c)\hat{L}\frac{\bar{P}_1}{\bar{P}_1 + \bar{P}_2} > 0$$

player C can find η, $\eta > 0$ (perhaps η is very small), which enables $\ddot{\pi} - \dot{\pi} > 0$. Now consider $\hat{L} > \bar{P}_1$; CP will not change and the profit of player C is

$$\ddot{\pi} = \sum_{j \in C, p_j < \rho^*} (\rho^* - c)\bar{P}_j + (\rho^* - c)\bar{P}_1 \tag{6.36}$$

The difference between $\dot{\pi}$ and $\ddot{\pi}$ is

$$\ddot{\pi} - \dot{\pi} = (\rho^* - c)\bar{P}_1\frac{\bar{P}_1 + \bar{P}_2 - \hat{L}}{\bar{P}_1 + \bar{P}_2} > 0 \tag{6.37}$$

The above calculation shows that player C has motivation to change the bid price of generator 1, which makes the price deviate from ρ^*. If player C has multiple generators participating in the auction, obviously an identical conclusion can be drawn. The same concept applies to player D. The proof is complete.

The premise is as follows. If an electricity price is determined simultaneously by a generator of a supplier and its rival's generator, the supplier should bid a price lower than the rival's price. As a result, the supplier would increase its output although the market CP would decrease. The supplier would gain more than it would lose. Now, let us discuss one of the main conclusions of this section.

Proposition 1 (equilibrium under tight capacity constraints)—Suppose there exists a player A such that

$$\sum_{j \notin A} \bar{P}_j < D_P$$

Then the game possesses an equilibrium at which $\rho^* = \bar{c}$. *Furthermore,* $\rho^* = \bar{c}$ *uniquely (Gan et al. 2005).*

Proof—Consider the following configuration. The bid price of all the generators of supplier A is $\bar{p}$ and the bid price of its rivals is c. The configuration is an equilibrium because (1) the rivals of supplier A have no motivation to bid prices deviating from c because their generators are all accepted, the CP is the highest possible, and all have gained maximum profits; and (2) supplier A has no motivation to deviate from its strategy. The reason is that if the bid price of supplier A is between c and $\bar{p}$, the profit of supplier A will decrease. If the bid price of supplier A is c, the profit of supplier A is zero—less than the profit under the condition

$$\left(\bar{p} - c \right) \left(D_P - \sum_{j \notin A} \bar{P}_j \right) > 0$$

We now prove the uniqueness of the proposition. Apparently no equilibrium at which $\rho = c$ exists. Suppose $c < \rho < \bar{p}$ and consider two possible configurations: (1) the clearing price is determined by an individual supplier and (2) the clearing price is determined simultaneously by several suppliers. Condition (1) has no equilibrium points because it has motivation to raise the price no matter which generator sets the clearing price. We use Lemma 2 to prove that condition (2) is not at equilibrium and the proof is complete.

The statement that "$\rho^* = \bar{p}$ *uniquely*" does not necessarily mean that the equilibrium is unique. For example, when

$$\sum_j \bar{P}_j = D_P$$

multiple equilibrium points can exist, but the clearing price of all the equilibrium points equals $\rho^* = \bar{p}$.

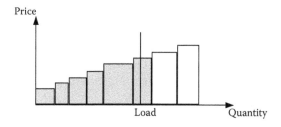

Figure 6.3 Competition under weak capacity constraints.

Now we study market performance under a weak capacity constraint under which a large supplier (shaded generators in Figure 6.3) may not have an incentive to bid a very high price. The following proposition describes equilibrium under weak capacity constraints.

Proposition 2 (equilibrium under weak capacity constraints)—I is a set for all the suppliers. If for each i, i ∈ I,

$$\sum_{j \notin i} \bar{P}_j > D_P$$

the equilibrium point ρ > c does not exist. In other words, if an equilibrium point exists, ρ = c at the equilibrium point (Gan et al. 2005).

Proof—Suppose an equilibrium point ρ > c exists and two possible configurations are (1) the clearing price is determined by an individual supplier and (2) the clearing price is determined simultaneously by several suppliers. To show that configuration (1) is not at equilibrium, recall the hypothesis

$$\sum_{j \notin A} \bar{P}_j > D_P$$

for every player *A, A ∈ I.* Therefore unaccepted generator(s) must be owned by the price-setting player's rival and have the incentive to undercut the price-setting generator because ρ > c. To prove that configuration (2) is not at equilibrium, readers are referred to Lemma 2. The proof is complete.

The result includes the classic Bertrand paradigm as a special case. Consider a market with two sufficiently large generators. At the equilibrium point of the market, the CP is equal to the marginal cost. As a more general example, consider a market with three generators (A, B, C). Each is owned independently and none can serve the load alone. Two of them can always serve the load and the clearing price of the market at equilibrium is equal to marginal cost *c.*

We end this section by discussing some of the drawbacks of the above model, which is, however, a simultaneous, perfect information, infinite, and discontinuous example. We have not described the equilibrium of the proposed model completely. We only describe the two extreme conditions of tight and weak generation capacity constraints. Another drawback of the model is that it is static while electricity market auctions in reality are dynamic and repetitive. Finally, it does not consider many important market features, for example, transmission rights, reserves, and generator characteristics.

6.4 Market Power Analysis

6.4.1 Market Power Measurement Indices

Traditionally, the Herfindahl–Hirschman index (HHI), entropy coefficient (EC), and Lerner index (LI) are used to measure market power. Both HHI and EC calculations are based on market shares. The indices can be formulated as Equations 6.38 and 6.39, respectively.

$$HHI = 10000 \sum_{i=1}^{N} s_i^2 \tag{6.38}$$

$$EC = -\sum_{i=1}^{N} s_i \ln(s_i) \tag{6.39}$$

where, s_i indicates the market share of participant i and N is the number of participants in the market. A high HHI represents a high market concentration while a high EC represents a low market concentration; for example, HHI = 10,000 and EC = 0 indicate a monopoly structure.

For example, two suppliers exist in a market and each supplier's market share is 50%, so HHI = $50^2 + 50^2$ = 5,000. Economists argue that market power probably exists if HHI exceeds 2,000. Unlike HHI and EC, LI can be formulated as

$$LI = \frac{P - MC}{P} \tag{6.40}$$

where P is the market price and MC is the marginal cost. LI reflects the degree of price uplift that belongs to ex post methods and does not appear ideal for electricity market application. Conversely, HHI and EC, which are based on market shares, are difficult to incorporate into the peculiarities of power systems.

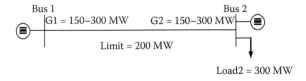

Figure 6.4   Generator G2 has market power.

6.4.2 Market Power Analysis

Consider the example in Figure 6.4. The figure clearly shows that generator G2 has a must-run generation of $300 - 200 = 100$ MW and therefore has market power. In fact, the price competition model described in the preceding section gives a theoretical foundation for calculating market power. Based on the model, we defined a numerical index (must-run ratio) that may be used to measure market players' abilities to manipulate prices. For the ith supplier, we calculate the index (Gan and Bourcier 2002) as follows:

$$MRR = \frac{D_P - \sum_{j \notin i} \bar{\mathbf{P}}_j}{\sum_{j \in i} \bar{\mathbf{P}}_j}, \; 0 \leq MRR \leq 1 \tag{6.41}$$

This numerical index reflects the ratio of must-run generation to total available generation of a generator. A salient feature is that these indices reflect ex ante the price-controlling *potentials* of individual suppliers. In the California electricity market, the Residual Supply Index is used to calculate the market power of suppliers. Its underlying principle is similar to that of the must-run ratio.

Although market power analysis may be difficult and sometimes controversial for market power evaluation, it is essential for successfully building and operating an electricity market. This applies to both active market power analysis and reactive market power analysis (Feng et al. 2008).

6.5 Electricity Market Experiments

The British experience demonstrates that the actual performance of an auction may be difficult to predict. One reason is that the game theoretic and other mathematical models may have multiple solutions. Therefore, experiments are employed to study economic behaviors (Backerman et al. 1997). Here we introduce experimental results the authors obtained at Zhejiang University in an attempt to shed light on this interesting subject. Serious readers should consult, for example, Chen and Wang (2007) and Zimmerman et al. (1999) for more detailed coverage.

In November 2002, about 50 graduate students taking an electricity market course in the Department of Electrical Engineering of Zhejiang University were selected to participate in two experiments. During the experiments, we learned that some of the students had not studied the principle of Nash equilibrium.

Experiment 1—a power system consisting of 4 generating units supplies a 1000 MW load. Each unit has a capacity of 300 MW and their marginal costs are identical. The market rules require that the minimum bid price is $0/MW and the price ceiling is $100/MW.

Experiment 2—A power system consisting of 4 generating units supplies a 1000 MW load. Each unit has a capacity of 400 MW, and their marginal costs are identical. The market rules require that the minimum bid price is $0/MW and the price ceiling is $100/MW.

According to the market analysis model, the price at the market equilibrium point in Experiment 1 equals the price ceiling, i.e., $100/MW. In Experiment 2, the price at the market equilibrium point equals the minimum price, i.e., $0/MW.

Most of the students in Experiment 1 did not select Nash equilibrium strategies. However, in Experiment 2, more students bid a lower (Nash) price. These results are partly in accord with the Nash results. The discrepancy, based on our experience, arises from the multiplicity of Nash equilibrium points.

Appendix 6A: Fixed Point Theory and the Existence of Market Equilibrium

Most (if not all) of the time, it is very difficult to calculate the Nash equilibrium accurately for an infinite game. As a remedy, we may wish to ensure the existence of Nash equilibrium. The fixed point theory for a multi-valued function is needed. A commonly used theorem for a multi-valued function is the celebrated Kakutani fixed point theorem. This section discusses a result that reflects the flavor of this theorem.

Lemma 3 (existence of Nash equilibrium [Gan and Shen 2004])—For the price competition models introduced in this chapter for a two-generator market, if the payoff functions $\pi_k(\mathbf{p})$, $k = 1,2$, are continuous in $\mathbf{s}_k$, the auction game has an equilibrium.

Proof—Notice that as $\mathbf{p}_k$ increases, $\pi_k(\mathbf{p})$ either decreases at a point or is non-decreasing. Since the additional hypothesis $\pi_k(\mathbf{p})$ is continuous, the pay-off function can only be non-decreasing. This indicates that $\pi_k(\mathbf{p})$, $k = 1,2$, is quasi-concave in $\mathbf{p}_k$, inferring that the reaction correspondence $\mathbf{r}(\mathbf{p})$, is *convex valued*.

The set expressed in Equation 6.31 is closed and the pay-off function $\pi_k(\mathbf{p})$ is continuous in $\mathbf{p}_k$. Under the Mas-Colell et al. (1995) theorem of maximum, the reaction correspondence $\mathbf{r}(\mathbf{p})$ is non-empty and *upper semi-continuous*. Apparently, $\mathbf{r}(\mathbf{p})$ maps the compact convex set $\left\{\mathbf{p} : \mathbf{c} \leq \mathbf{p} \leq \bar{\mathbf{p}}\right\}$ into *itself.* All the conditions required by the Kakutani fixed point theorem are satisfied, so $\mathbf{r}(\mathbf{p})$ has a fixed point. The proof is complete.

References

C. D. Aliprantis and S. K. Chakrabarti. 2000. *Games and Decision Making.* New York: Oxford University Press.

S. R. Backerman, M. J. Denton, S. J. Rassenti et al. 1997. Market Power in a Deregulated Electrical Utility Industry: An Experimental Study. Tucson: University of Arizona Economic Science Laboratory. Working Paper.

J. B. Cardell, C. C. Hitt, and W. W. Hogan. 1997. Market power and strategic interaction in electricity networks. *Recourse and Energy Economics,* 19, 109–137.

H. Chen and X. Wang. 2007. Strategic behavior and equilibrium in experimental oligopolistic electricity markets. *IEEE Transactions on Power Systems,* 22.

D. Feng, J. Zhong, and D. Gan. 2008. Measuring reactive market power using must-run indices. *IEEE Transactions on Power Systems,* 23, 755–765.

D. Gan and D. V. Bourcier. 2002. Locational market power screen and congestion management: experience and suggestions. *IEEE Transactions on Power Systems,* 17, 180–185.

D. Gan and C. Shen. 2004. A price competition model for power and reserve market auctions. *Electrical Power Systems Research,* 70, 187–193.

D. Gan, J. Wang, and V. B. Donald. 2005. An auction game model for pool-based electricity markets. *International Journal of Electrical Power and Energy Systems,* 27, 480–487.

R. J. Green and D. M. Newbery. 1992. Competition in British electricity spot market. *Journal of Political Economy,* 100, 929–953.

G. A. Jehle and P. J. Reny. 2000. *Advanced Microeconomic Theory.* Boston: Addison Wesley.

A. Mas-Colell, M. Whinston, and J. Green. 1995. *Microeconomic Theory.* Oxford, U.K.: Oxford University Press.

J. D. Weber and T. J. Overbye. 1999. A two-level optimization problem for analysis of market bidding strategies. *IEEE Power Engineering Society Summer Meeting,* pp. 682–687.

F. Wen and A. K. David. 2001. Optimal bidding strategies and modeling of imperfect information among competitive generators. *IEEE Transactions on Power Systems,* 16, 15–21.

R. Zimmerman, R. Thomas, D. Gan et al. 1999. A Web-based platform for experimental investigation of electric power auctions. *Decision Support Systems,* 24, 193–205.

Bibliography

F. Alvarado. 1999. The stability of power system markets. *IEEE Transactions on Power Systems,* 14, 505–511.

X. Bai, S. M. Shahidehpour, V. C. Ramesh et al. 1997. Transmission analysis by Nash game method. *IEEE Transactions on Power Systems,* 12, 1046–1052.

J. F. Bard. 1998. *Practical Bilevel Optimization*. Boston: Kluwer.

L. B. Cunningham, R. Baldick, and M. L. Baughman. 2002. An empirical study of applied game theory: transmission-constrained Cournot behavior. *IEEE Transactions on Power Systems*, 17, 166–172.

Y. He, Y. H. Song, and X. F. Wang. 2002. Bidding strategies based on bid sensitivities in generation auction markets. *IEEE Proceedings on Generation, Transmission, and Distribution*, 149, 21–26.

B. F. Hobbs, C. Metzler, and J. Pang. 2000. Strategic gaming analysis for electric power networks: an MPEC approach. *IEEE Transactions on Power Systems*, 15, 638–645.

P. L. Joskow. 1997. Restructuring, competition, and regulatory reform in the U.S. electricity sector. *Journal of Economic Perspectives,* 11, 119–138.

D. Luenberger. 1995. *Microeconomic Theory*. New York: McGraw Hill.

M. Shahidehpour, H. Yamin, and Z. Li. 2002. *Market Operations in Electric Power Systems*. New York: John Wiley & Sons.

N. H. M. von der Fehr and D. Harbord. 1993. Spot market competition in the U.K. electricity industry. *Economic Journal*, 103, 531–546.

Chapter 7

Price Forecast and Risk Management

Electricity prices are well known for their volatility. This chapter discusses management of financial risks resulting from variable electricity prices. Generation companies, utility companies, and other electricity sellers and buyers face the need to maximize profits and minimize risks arising from highly volatile electricity prices. One efficient way to manage risk is diversification, i.e., trading in different markets with different counterparts and participating in various types of energy and ancillary service contracts. This chapter covers portfolio management techniques and their effectiveness in real trading decision making. We will also discuss the techniques of price forecasting that serve as the bases of risk management.

7.1 Forecasting Electricity Prices

The characteristics of electricity are distinctly different from those of other commodities. For example, electricity is not storable and requires expensive unit start-up investments. Electricity price movement shows more volatility. For this reason, electricity price forecasting is extremely important for all market players. Short-term price forecasting is intended to estimate electricity prices for a few days into the future, while mid- and long-term price forecasting focuses on estimates covering several months.

The results of short-term price forecasting can be used by a transmission company for several purposes, including scheduling short-term generator outages and designing load response programs. Generation companies also use forecasting. Mid- and long-term forecasting results serve many purposes, for example, assessing grid company risks, facilitating optimal transmission planning, evaluating generation side market power and developing mitigation procedures, and aiding negotiations with transmission companies to ensure local voltage support. Based on market needs, several methods are capable of forecasting market prices.

Artificial neural networks (ANNs) and time series models are used widely in short-, mid-, and long-term electricity price forecasting. Market power, as discussed in Chapter 6, is a critically important factor. In the sequel we describe some price forecasting results using ANN and time series methods, with considerations to market power.

7.1.1 Short-Term Price Forecasting

Several approaches to short-term price forecasting have been reported. The similar day price forecasting method and time series analysis have long been used for short-term load forecasting. It is our understanding that these approaches do not fit well in the context of price forecasting because they do not account for impacts such as planned outages and market power. ANNs appear to be natural choices for short-term price forecasting because of their ability to model complex issues determining the behaviors of market players and ensuing market clearing prices.

To achieve a successful ANN application, input variables for constructing the ANN must be chosen judiciously. The most important decision is whether different ANNs should be constructed for different time periods. We believe that this is unnecessary because market clearing prices are determined by the bidding strategies of suppliers. It is natural to believe that a supplier mainly considers forecast load along with the costs and capabilities of its generators and those of its rivals. These factors may show certain degrees of periodicity but periodicity is not an inherent characteristic.

A simple three-layer neural network is described in this chapter. Its three input variables are forecast load, MRR, which is defined in 6.41 and the market price of the trading period of the previous day. The output of the network is the forecast price. Figure 7.1 ilustrates the structure of the suggested neural network.

To improve price forecasting accuracy, a threshold of MRR is set up. If the MRR of a trading period exceeds the threshold, the output of the neural network is abandoned and the price cap is substituted as the forecasting output. In fact, the suggested price forecasting approach combines the spirits of neural networks and game theory. Figure 7.2 is a flowchart depicting the overall forecasting method.

The forecast results using the proposed approach appear in Figure 7.3. To allow a comparison, the forecast results without a game theoretical adjustment appear in Figure 7.4. The dip of the actual price at period 23 (11:30 a.m.) can be explained. The load curve has a 1000 MW dip in that trading period. To ensure that production is allocated according to the contract for difference and to avoid expensive unit start-up costs, many suppliers bid negative prices for that period to keep their units running. In addition, the actual settlement price is zero if the clearing price is below

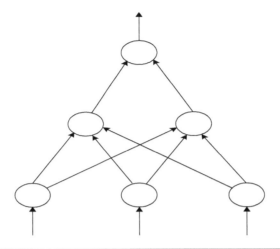

Figure 7.1   Structure of the suggested neural network.

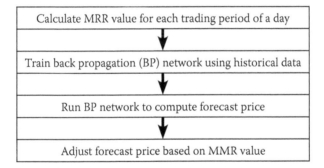

Figure 7.2   Flowchart of short-term price forecasting algorithm.

zero based on market rules. This is why the clearing price for that period is often below zero. It is based on the willingness of generators. To measure the relative errors of the software, the following numerical index is calculated:

$$\sigma_{MAPE} = \frac{1}{N} \sum_{i=1}^{N} \frac{\left| p_{MCP,i} - p'_{MCP,i} \right|}{p_{MEAN}} \quad (7.1)$$

where $p_{MCP,i}$ and $p'_{MCP,i}$ are actual and forecast Market Clearing Prices (MCPs), respectively; p_{MEAN} is the average daily price; N is the number of forecast periods. The absolute errors are defined as

$$\sigma_{MAE} = \frac{1}{N} \sum_{i=1}^{N} \left| p_{MCP,i} - p'_{MCP,i} \right| \quad (7.2)$$

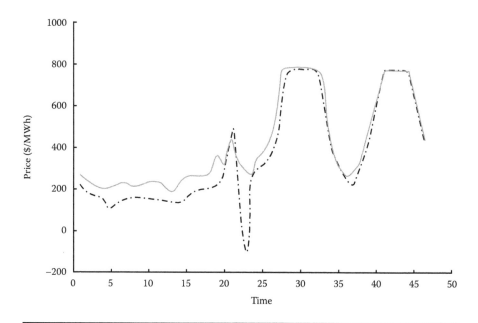

Figure 7.3   Actual prices versus forecast prices with adjustment of July 12, 2002.

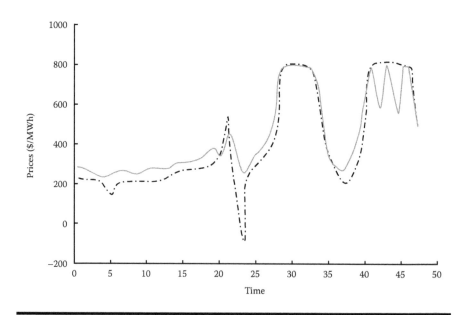

Figure 7.4   Actual prices versus forecast prices without adjustment of July 12, 2002.

Table 7.1   Forecast Errors

Forecast Day	σ_{MAE} ($/MWh)	σ_{MAPE} (%)
July 9	40.782	19.145
July 10	39.901	13.765
July 11	59.440	17.382
July 12	59.587	15.291
July 13	51.097	9.7132
July 14	28.898	16.446
July 15	72.985	16.893

A week in summer that yields significant outliers was selected to test. During the week, the MCPs covered a wide range (−659.93 to 819.80). Furthermore, as many as four price spikes can occur within a day in the period. Table 7.1 summarizes the error measurements.

Bunn (2000) noted that the relative errors in price forecasting cannot be within 10%. The results summarized in Table 7.1 seem to support this assertion. It is worth noting that the precision of the introduced price forecasting approach can be greater by selecting higher if a period during which market clearing prices are relatively stable is selected.

We complete this section by mentioning that the neural network described requires only a few input variables. This is because the MRR variable effectively captures market performance.

7.1.2 Mid- and Long-Term Forecasting Using Linear Regression

Many factors such as fuel costs, bank rates, generator and transmission outages, weather, and market share changes affect mid- and long-term electricity market prices. Many of these factors involve uncertainties and thus may not be predicted accurately. Among several approaches considered, fuzzy set theory seems suitable but does not provide the quantitative information needed for subsequent risk evaluation. A neural network is rejected because it does not yield a confidence interval—a basic requirement of analysis. Linear regression was chosen as the price forecasting approach.

Assume $x_1, x_2, ..., x_p$ ($p > 1$) are independent random variables. The output of the linear regression model y is calculated as

$$y = b_0 + b_1 x_1 + ... + b_p x_p + \varepsilon \tag{7.3}$$

where $N(0,\sigma^2)$ denotes the usual Gauss distribution; $b_0, b_1, ..., b_p$, σ^2 are unknown constants, and ε is a random variable with Gauss distribution, i.e., $\varepsilon \sim N(0,\sigma^2)$.

The most important task is to choose the input variables $x_0, x_1, \ldots, x_p$ to achieve the best price forecast. After a careful analysis, the selected input variables were forecast load, MRR, generation cost, and previous week market prices. Again, we found that MRR was an important input variable for the linear regression model.

A week is a basic forecast period. The market clearing price data for 45 weeks from Spring 2001 to Autumn 2002 were collected. We used data for 40 weeks to construct the linear regression model and data for 5 weeks to test the regression model. The results, including confidence intervals, are summarized in Table 7.2.

While we can calculate the average error as 14.47%, the actual prices all fell into the confidence intervals. Price forecast results with such fairly high quality are very useful for predicting conditions in electricity markets. A correlation analysis was performed to determine which factors most affected mid- and long-term prices. Figure 7.5 summarizes the results.

We can conclude from the figure that MRR has the most significant impact on mid- and long-term electricity market prices and is thus an excellent index for describing market performance.

Table 7.2   Medium- and Long-Term Electricity Price Forecast Results

Week	Confidence Interval	Market Price	Error %
1	178.0905 ± 58.1055	149.07	19.468
2	332.4460 ± 80.1684	286.83	15.903
3	288.0426 ± 40.9207	251.71	14.434
4	417.3471 ± 67.9940	383.01	8.965
5	175.1862 ± 66.7511	202.66	13.557

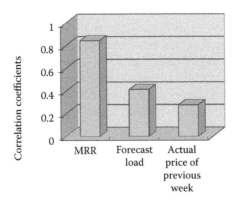

Figure 7.5   Correlation coefficients.

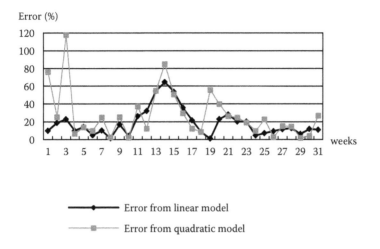

Figure 7.6   Error comparison of linear and quadratic regression models.

We complete this section by mentioning that the authors constructed a quadratic regression model and tested it against actual market data. As a comparison, both models were tested under worst case and ordinary case circumstances. In the worst case, forecasting results based on the linear regression model revealed more than 25% errors on average. The ordinary case yielded errors below 20% on average based on a linear regression model. The forecasting results for the 31 weeks between April 2002 and December 2002 covering the worst and ordinary cases were collected. Figure 7.6 illustrates the comparison. The quadratic regression model displayed no superiority over the linear regression model. The figure shows that the linear model is preferable from a practical view.

7.2 Managing Price Risk through Trading Portfolio Optimization

In a deregulated environment, market participants (generation companies, utility companies, and other sellers and buyers) are confronted with electricity price volatilities and other uncertainties. Their objective, of course, is to maximize profits while minimizing related risks.

Diversification is one of the most efficient strategies for reducing nonsystematic risks in stock markets. For an electricity market participant, one practical way to benefit from diversification is participating in various types of trading activities (forward contracts, contracts for differences, interruptible contracts, and day-ahead energy and reserve bidding) to construct an efficient portfolio.

This section discusses three categories of methods for establishing an optimal trading portfolio in an electric industry: (1) the mean variance method that yields an analytical solution, (2) the (conditional) value at risk (VaR/CVaR) method that considers the constraints or objectives of the traditional profit maximization problem, and (3) simulation-based method that can incorporate complex risk attitudes of participants, whose solution accuracy depends on the risk attitude model and the simulation time.

For a detailed explanation of Method (1), readers can refer to Feng et al. (2007). For more information about Method (2), readers can refer to Carrión et al. (2007) and Conejo et al. (2008). A case of Method (3) is demonstrated in Section 7.2.1. Section 7.2.2 covers the comparison between the methods.

7.2.1 A Simulation-Based Method

We start our discussion with the expected utility theory (EUT) that is widely accepted in decision-making studies under uncertainty (von Neumann and Morganstern 1947). EUT assumes that an investor's objective is to maximize the expected value of its utility function. The supplier trading portfolio decision can be formulated as

$$\max_{q_1, q_2, \ldots, q_M} \mathrm{EU}(\tilde{V}) \tag{7.4}$$

$$s.t. \quad \sum_{m=1}^{M} b_m \cdot q_m = P \tag{7.5}$$

$$b_m \cdot \underline{q_m} \le q_m \le b_m \cdot \overline{q_m}, \, m = 1, \ldots, M \tag{7.6}$$

$$b_m = \{0,1\}, \, m = 1, \ldots, M \tag{7.7}$$

In this optimization formulation, the objective function $\mathrm{EU}(\tilde{V})$ represents a generator's expected utility, where $\tilde{V}$ denotes the trading revenue and q_m denotes the quantity of electricity sold through the mth trading type (asset). Equation 7.5 represents the generation capacity constraint where P denotes the output of the generator and b_m is a binary variable to determine whether asset m is chosen. Equation 7.6 represents the trading constraints where $\underline{q_m}$ and $\overline{q_m}$ denote the lower and upper limits of the mth available trading type.

The model above formulates the trading allocation (here-and-now) decision while the production decision determining recourse actions is simplified. We assume that the generator acts as a price taker in the spot market and continuously operates at its full capacity in the trading period. This is close to the situation of base-load generators such as large coal-fired plants and nuclear power plants. If a production decision is considered, generator output should depend on operation intervals. Operation issues such as start-up and ramping constraints should be incorporated.

We now derive the specific formulation of $EU(\tilde{V})$. Since trading revenue $\tilde{V}$ depends on the trading portfolio and asset prices, $EU(\tilde{V})$ can be formulated as

$$EU(\tilde{V}) = E\left\{U[\sum_{m=1}^{M}\sum_{t=1}^{T}\tilde{p}_{m,t} \cdot q_m]\right\} \tag{7.8}$$

where $\tilde{p}_{m,t}$ denotes the price of asset m at trading interval t and $U(\bullet)$ denotes investors' utility functions. Usually $U(\bullet)$ is assumed to be nondecreasing and concave (risk-averse assumption). In financial literature, investors are usually assumed as risk-averse because of the statistics generated by questionnaire investigations and psycho-economic experiments (the most famous is Bernoulli's St. Petersburg experiment) and also because of market imperfections and frictions. Electricity markets are widely recognized as incomplete and risk-averse assumptions are thus suitable for most power companies.

Since $\tilde{p}_{m,t}$ is a stochastic variable, $\tilde{V}$ is also stochastic. The mean variance-based analytical (MV-A) model assumes that the trading revenue $\tilde{V}$ of all feasible portfolios is normally distributed. Therefore a utility can be described only by the mean and the variance of portfolio return in the simplest linear combination form:

$$EU(\tilde{V}) = E(\tilde{V}) - \alpha \cdot D(\tilde{V}) \tag{7.9}$$

where α is a weighting factor representing the decision maker's attitude toward risk. $E(\tilde{V})$ and $D(\tilde{V})$ denote the expectation and variance of trading revenues, respectively.

However, the normality of $\tilde{V}$ (i.e.,

$$\sum_{m=1}^{M}\sum_{t=1}^{T}\tilde{p}_{m,t} \cdot q_m$$

for any feasible bundle of q_ms) depends on a group of factors such as the number of available trading types M, the diversification of the trading portfolio (q_m), the length of trading period T, and the correlation and distribution of asset prices $\tilde{p}_{m,t}$.

For example, if a the trading portfolio consists of spot energy, spot reserves, and fixed-price forwards whose distributions are close to normal (real data usually exhibit asymmetric and fat-tail or high-peak attributes), the MV-A model is a proper approximation of expected utility. Conversely, if a trading portfolio is dominated by nonlinear assets such as optional forwards, one-way CfD, and other heavily asymmetric assets, the normal distribution assumption is inappropriate and may create an inaccurate portfolio analysis, as the numerical study in this section will demonstrate.

The motivation is to propose a method that can accurately consider the distribution of normal and nonnormal assets. We start from a general form of $EU(\tilde{V})$. When the joint distribution of asset prices is known, $EU(\tilde{V})$ can be formulated as

$$EU(\tilde{V}) = \int_{-\infty}^{+\infty} \cdots \int_{-\infty}^{+\infty} U\left[\sum_{m=1}^{M}\sum_{t=1}^{T} \tilde{p}_{m,t} \cdot q_m\right] \tag{7.10}$$

$$\cdot D(\tilde{p}_{1,1}, \ldots, \tilde{p}_{1,T}, \ldots, \tilde{p}_{M,T}) d\,\tilde{p}_{1,1}, \ldots, d\,\tilde{p}_{1,T}, \ldots, d\,\tilde{p}_{M,T}$$

where $D(\tilde{p}_{1,1}, \ldots, \tilde{p}_{1,T}, \ldots, \tilde{p}_{M,T})$ denotes the joint probability density functions of stochastic variables $\tilde{p}_{1,1}, \ldots, \tilde{p}_{1,T}, \ldots, \tilde{p}_{M,T}$. If we further assume that all the stochastic variables $\tilde{p}_{1,1}, \ldots, \tilde{p}_{1,T}, \ldots, \tilde{p}_{M,T}$ are independent of each other, we have

$$EU(\tilde{V}) = \int_{-\infty}^{+\infty} \cdots \int_{-\infty}^{+\infty} U\left[\sum_{m=1}^{M}\sum_{t=1}^{T} \tilde{p}_{m,t} \cdot q_m\right] \tag{7.11}$$

$$\cdot D_{1,1}(\tilde{p}_{1,1}), \ldots, \cdot D_{1,T}(\tilde{p}_{1,T}), \ldots, \cdot D_{M,T}(\tilde{p}_{M,T}) d\,\tilde{p}_{1,1}, \ldots, d\,\tilde{p}_{1,T}, \ldots, d\,\tilde{p}_{M,T}$$

where $D_{m,t}(\tilde{p}_{m,t})$ denotes the probability density function of $\tilde{p}_{m,t}$.[*] In electricity markets, the assumption that all asset prices are independent is rare because market operation demonstrates that prices of some assets (day-head and real-time markets) are highly correlated. The independence assumption also cannot hold for assets such as one-way CfD whose prices depend partially on spot price. Electricity futures prices are also correlated theoretically and practically with spot prices.

Even if we can obtain a price distribution of each asset at each trading interval in the decision period, $D(\tilde{p}_{1,1}, \ldots, \tilde{p}_{1,T}, \ldots, \tilde{p}_{M,T})$ it is still difficult to unbundle and thus difficult to formulate explicitly. This leads to the inaccessibility of the analytical form of $EU(\tilde{V})$. Therefore, the gradient vector or other analytical information necessary for a traditional programming approach is unavailable. For such a non-differentiable optimization problem with stochastic parameters that is difficult to solve by classical programming methods, a simulation embedded artificial intelligence (AI) method is a natural choice.

---

[*] The density functions can be drawn, for example, from stochastic analysis of historic prices.

Monte Carlo simulation makes the proposed method able to obtain price samples without knowing the joint distribution of assets' prices. A genetic algorithm (GA) can optimize without analytical information such as the gradient vector. The proposed methodology embeds Monte Carlo simulation into the fitness evaluation module of a GA. The algorithm can handle general supplier trading decisions and overcome the dilemmas of extending trading types and the constraints of distribution assumptions. However, we must note that although a GA is applied here, other AI search and optimization methods such as simulated annealing, beam search, and random optimization can replace a GA in the proposed EUT-S framework. Figure 7.7 is a flow chart of the solution procedure and interactions of major modules of the proposed EUT-S algorithm. The functioning process can be implemented in nine steps:

1. Initialize the program, set the parameters (number of individuals, number of generations, probability of genetic operations, etc.) and randomly generate a number of feasible portfolios.
2. Evaluate the fitness of all the feasible portfolios. The fitness value is calculated via a Monte Carlo approach simulated from historical asset price data. First, assume the decision period consists of T trading intervals. For each trading interval, randomly select the asset price from the reference sample pool* and calculate the utility of each sample by

$$U\left[\sum_{t=1}^{T}\sum_{m=1}^{M}(\hat{p}_{m,t} \cdot q_{m,t})\right] \tag{7.12}$$

$\hat{p}_{m,t}$ denotes the randomly selected[†] price from the sample pool of the mth asset at the tth trading interval. Repeating the sample selection and utility calculation, we obtain a series of sample utilities. The fitness of the ith portfolio is set as the average of these sample utilities.

---

* The reference sample pool can be composed of the asset prices of similar trading intervals in a history. In the PJM system in the U.S., each hour has a spot price and the trading interval is regarded as 1 hour. Similar trading intervals can be the same hour of the day, the same hour 10 days previously, and the same hour 20 days later. Our simulation also considers day type (weekday or weekend). The similar trading intervals of weekdays will exclude the historical weekend days and vice versa. Since the historical data used in this section cover 1999 to 2004, the fluctuations among years should also be considered. Historical data are adjusted by a ratio to capture the long-term price trend. The adjusting ratio is calculated by a curve-fitting approach on a yearly basis.
† A selection probability is assigned to each sample in the pool according to its timeline. Recent data are assigned with greater selection probability.

3. Reorder the feasible portfolios according to their fitness values. Check whether the program satisfies the stopping criterion.

   Two stopping criteria are employed in our program: (1) the improvement in the best fitness value is less than tolerance for 50 generations; (2) the program reaches the maximum iteration number. The program will stop if either criterion is satisfied.

4. Select the best N portfolios and encode them from variable space (decimal) into gene space (binary) to form the parent generation in the gene space.

5. Let the parent generation run genetic operations (reproduction, crossover, and mutation). First we reproduce the parent generation based on the fitness of each individual within it. Some of the reproduced individuals participate in the crossover operation and some participate in the mutation operation based on the crossover and mutation probabilities. Some of the best individuals directly constitute the child generation. After the genetic operation, a new (child) generation of individuals is born. Figure 7.7 is the flow chart.

6. Decode all individuals in the offspring generation from gene space into variable space.

7. Test the feasibility of all the portfolios and eliminate the infeasible ones.

8. Repeat steps 2 through 7 until the program stops at step 3.

9. Choose the best portfolio in the current generation (the generation in which the program stops) as the output result.

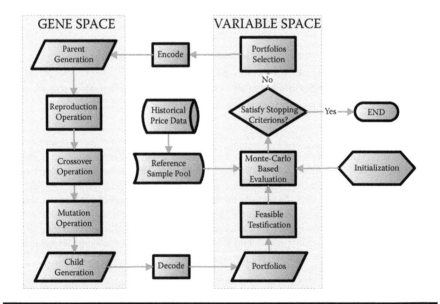

Figure 7.7   Algorithm flow chart.

In this section, based on EUT model analysis, an algorithm of the simulation-based method is presented to solve the generalized supplier trading portfolio decision. In Section 7.2.2, we will test this particular simulation method with real market data-based examples and compare its performance with the standard MV-A method.

7.2.2 Application and Comparison

In practical application, decision makers should consider computation time, model and solution accuracy, and general usability of an algorithm. The EUT-S method can consider the nonnormality of electricity assets properly and thus outperform the MV-A method after a number of iterations—at a cost of more simulation time. The simplicity of the MV-A method makes it much faster than the EUT-S method.

The MV-A is also more convenient for parameter setting because it needs only one parameter (degree of risk aversion) to express a decision maker's risk return attitude. The EUT-S method adopts EUT to portray a decision maker's risk-return profile. The utility function often needs three or four parameters (three for quadratic, exponent, and logarithmic utility functions and four for the general linear risk tolerance [LRT] utility function). These different forms of utility functions enable the method to portray more diverse decision makers with diverse attitudes toward risk. Examples of risk attitudes are DARA (decreasing absolute risk aversion), DRRA (decreasing relative risk aversion), CARA (constant absolute risk aversion), and CRRA (constant relative risk aversion). The variety also makes the parameter settings more complicated.

In addition to the MV-A and EUT-S methods, VaR- and CVaR-based optimization (CVaR-O) methods are also widely used in portfolio decisions. VaR is not generally a coherent risk measure because it does not consider the sub-additivity property (Artzner et al. 1999). An immediate consequence is that VaR may discourage diversification.

CVaR is an improvement of VaR and is a coherent risk measure. CVaR does not require any specific distribution properties and is thus appropriate for handling nonnormal distributions. The easy parameter setting and small computation burden also contribute to the value of CVaR as a risk measure in electricity market research. CVaR is defined as the average of losses exceeding VaR. CVaR ignores the difference in higher moments above the VaR level.

The MV-A, CVaR-O, and EUT-S methods all have advantages and disadvantages when applied to trading portfolio decision problems. CVaR is an efficient risk measure for fast evaluation of nonnormal portfolios, for example, in multi-period portfolio analysis. The MV-A method performs well for normally distributed or almost normally distributed portfolio decisions, for example, in long-term, large-scale, well-diversified trading portfolio analysis. EUT-S is suitable where high level accuracy is required and computation time is sufficient.

In practical application, one important preliminary step is checking how well the randomness can be modeled by normal distribution. The three ways to check the normality of a random variable are (1) use graphical methods such as histograms and quantile–quantile (QQ) plots; (2) perform a back-of-the-envelope test; (3) calculate high order statistics (since the only two parameters of normal distribution are lower order statistics—first order moment mean and second order moment variance—nonzero higher order moments such as skewness and kurtosis indicate nonnormality and the larger the value, the heavier the nonnormality); (4) perform recent frequentist (D'Agostino's K-squared, Jarque-Bera, Anderson-Darling, Cramér-von Mises, Lilliefors normality, Shapiro-Wilk, Pearson's chi-squared, and Shapiro-Francia) tests and Bayesian tests.

To clarify the importance of a normality test in electricity trading analysis, we provide an example from Feng et al. (2010). Assume that a generating company owns a 300 MW base load unit located at the PJM western hub. The company is assumed to seek an optimal trading portfolio and its decisions are based on historical PJM market data from 1999 to 2004 (available on the PJM website).

Assume four trading settlements are available for the generator to sell electricity: a fixed-price contract (FC), a supplier-flexible CfD (CfDS), a consumer-flexible CfD (CfDC), and the day-ahead market (DA). The details of these four trading instruments are listed in Table 7.3.

CfDS and CfDC are types of one-way CfDS. A CfDC hedges the high spot prices for the consumer and the consumer pays the supplier a certain compensation. CfDS hedges the low spot prices for the supplier and the supplier pays the consumer a certain compensation. The prices of CfDS and CfDC are illustrated in Figure 7.8. When the spot price exceeds the CfDC threshold, the CfDC price will be CfDC *threshold plus compensation*. When the spot price is smaller than the CfDS threshold, the CfDS price will be CfDS *threshold plus (negative) compensation*.

To show an appropriate comparison with the MV-A method, the utility function of the proposed EUT-S method is assumed as

$$U(\tilde{v}) = 1.0068 - 1.0068 \cdot e^{-\alpha \cdot \tilde{v}} \qquad (7.13)$$

where $\tilde{v}$ is defined as $\tilde{v} = \tilde{V}/(T \cdot C(P))$ and represents a relative revenue level. The reasons an exponent function is chosen as the utility formulation are (1) an exponent function as an increasing and concave function is widely applied as the utility function; and (2) more importantly, an exponent utility function exhibits

Table 7.3 Trading Instruments

Asset	FC	DA	CfDC	CfDS
Trading price ($)	48	P_{DA}	min(P_{DA}, threshold)	max(P_{DA}, threshold)
Threshold ($)	N/A	N/A	60	50
Compensation ($)	N/A	N/A	1	–5

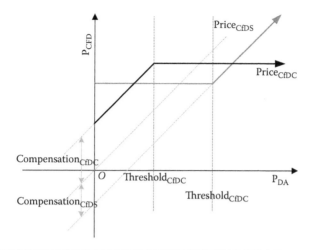

Figure 7.8   Prices of CfDC and CfDS.

Table 7.4   Portfolio Results and Higher Moments Magnitude (hourly)

Asset	FC	DA	CfDC	CfDS
MV-A Portfolio	40.78%	5.80%	53.01%	0.42%
EUT-S Portfolio	44.15%	42.41%	0.92%	12.52%
Skewness	NaN	2.314	−0.529	3.030
Kurtosis	NaN	10.257	2.121	13.735

constant absolute risk aversion (CARA) so that the absolute risk aversion degree will stay constant as $\tilde{v}$ varies, i.e., $-U''(\tilde{v})/U'(\tilde{v}) = \alpha$. When an exponent function is employed as the utility function, the EUT-S method is compared with the MV-A method on the same basis of risk preference.

In this study, performances of the EUT-S (1,000 generations of evolution) and the MV-A methods are compared for different trading periods (1 hour, 1 day, 1 week, 1 month, and 1 year). The results appear in Tables 7.4 through 7.8. The MV-A portfolio is calculated by the same method reported in Feng et al. (2007). The solutions shown in the tables indicate that the two approaches generate different optimal portfolios. Which is better? To compare the quality of the MV-A and EUT-S portfolios, a specific criterion is needed. We employ the CVaR adjusted expectation as the comparison criterion:

$$U_{CVaR}(\tilde{v}) = E(\tilde{v}) - \beta \cdot CVaR(\tilde{v}) \tag{7.14}$$

Table 7.5 Portfolio Results and Higher Moments Magnitude (daily)

Asset	FC	DA	CfDC	CfDS
MV-A Portfolio	41.11%	6.47%	50.69%	1.73%
EUT-S Portfolio	44.37%	25.43%	21.90%	8.30%
Skewness	NaN	1.209	−0.520	2.063
Kurtosis	NaN	5.338	2.697	8.461

Table 7.6 Portfolio Results and Higher Moments Magnitude (weekly)

Asset	FC	DA	CfDC	CfDS
MV-A Portfolio	45.78%	9.80%	38.34%	6.09%
EUT-S Portfolio	48.03%	21.43%	18.57%	11.96%
Skewness	NaN	0.552	−0.491	1.404
Kurtosis	NaN	3.226	2.854	4.747

Table 7.7 Portfolio Results and Higher Moments Magnitude (monthly)

Asset	FC	DA	CfDC	CfDS
MV-A Portfolio	50.44%	12.13%	28.67%	8.75%
EUT-S Portfolio	52.13%	15.43%	22.57%	9.86%
Skewness	NaN	0.254	−0.365	0.842
Kurtosis	NaN	1.503	2.908	3.556

Table 7.8 Portfolio Results and Higher Moments Magnitude (yearly)

Asset	FC	DA	CfDC	CfDS
MV-A Portfolio	52.78%	12.50%	25.34%	9.39%
EUT-S Portfolio	53.80%	13.43%	23.24%	9.53%
Skewness	NaN	0.300	−0.155	0.705
Kurtosis	NaN	1.017	2.927	2.934

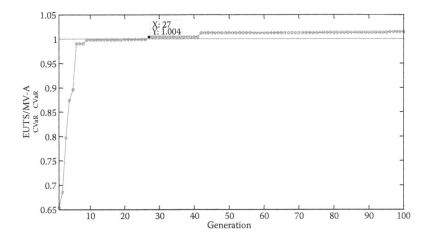

Figure. 7.9 Performance comparison of daily trading between EUT-S (first 100 generations) and MV-A.

where:

$$CVaR(\tilde{v}) = E\left(x \mid x < VaR(\tilde{v})\right) \tag{7.15}$$

$$VaR(\tilde{v}) = \sup\left(x \mid \text{Prob}(\tilde{v} < x) \leq 1 - c\right) \tag{7.16}$$

where $E(\bullet|\bullet)$ denotes the conditional expectation, $sup(\bullet)$ denotes the supremum, $\text{Prob}(\bullet)$ denotes conditional probability function, β denotes the CVaR adjusting factor (set according to risk aversion rate α), and c denotes the confidence level (set as 95%).

The U_{CVaR} of the best individual in each generation is calculated to test the performance of the evolution process. The results are shown as the open circles connected with the solid lines in Figure 7.9 and Figure 7.10. The y-axes in the figures are set as $U_{CVaR}(\tilde{v}^*_{EUT\text{-}S})/U_{CVaR}(\tilde{v}^*_{MV\text{-}A})$ to clearly show the U_{CVaR} difference between the EUT-S and MV-A portfolios. The figures indicate that although the performance of EUT-S will improve slowly or even remain the same for some generations, the general evolution trend is to produce individuals with better U_{CVaR}.

The dashed lines denote the level of the MV-A optimal portfolio. We can see that the EUT-S daily (monthly) portfolio outperforms the MV-A daily (monthly) portfolio after 26 (112) generations. The phenomenon of EUT-S outperformance generation is not a coincidence. It also occurs in the hourly, weekly, and yearly trading data. The results of average outperformance generation and simulation time of EUT-S are summarized in Table 7.9.*

---

* These simulations were performed by a prototype computer program based on MATLAB®. The hardware was a 2.53 GHz Intel E7200 (duo-core) processor with 2 GB RAM memory.

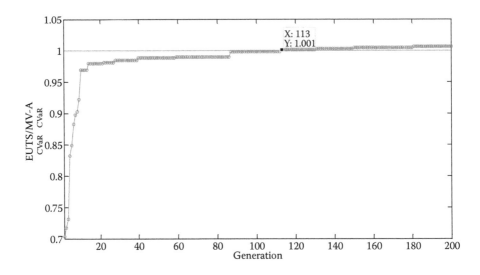

Figure 7.10 Performance comparison of monthly trading between EUT-S (first 200 generations) and MV-A.

Table 7.9 Average Outperformance Generation and Simulation Times for Various Trading Periods

Period	Hourly	Daily	Weekly	Monthly	Yearly
Generation	18.7	27.2	45.9	113.1	852.1
Time (s)	34.2	49.8	86.2	221.7	1783.9

EUT-S outperforms MV-A because the MV-A method considers only the first moment (mean) and second moment (variance) of the distribution of asset prices while the higher order moments are ignored. To gain an insight of the effect of the higher moments, the skewness (third standardized central moment) and the kurtosis (fourth standardized central moment) of the assets are calculated and listed in the bottom rows of Tables 7.4 through 7.8.

Note that DA and CfDS exhibit right-skewed (positive skewness) and high-peak/fat-tail (>3 kurtosis) characteristics. CfDC exhibits left-skewed (negative skewness) and low-peak/short-tail (<3 kurtosis) characteristics. The asymmetry and peak/tail characteristics can be viewed more clearly through the frequency distributions of the reference sample pools of these assets. In this chapter, we report only the frequency distribution of hourly and daily trading assets in Figures 7.11 through 7.16. Normal distribution curves are also shown for comparison purposes. We can clearly see the asymmetry and peak/tail characteristics in these figures.

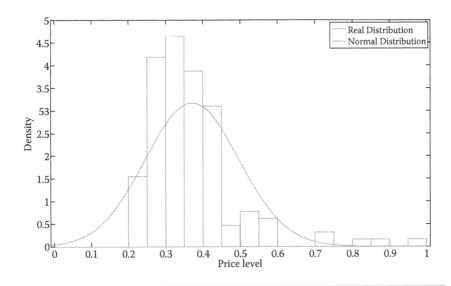

Figure 7.11 Frequency distribution of DA price (hourly).

The above statistical analyses and graphic illustrations show that the nonnormality of electricity trading assets is significant. The failure to consider nonnormality impairs the effectiveness of the MV-A approach and creates differences between the optimal portfolios calculated by the MV-A and EUT-S methods. This effect can be further revealed if we again review Tables 7.4 through 7.8. We can see that all the assets with positive skewness occupy a larger proportion of the EUT-S portfolio and all the assets with negative skewness occupy a smaller proportion of the EUT-S portfolio.

An important result we can note in Table 7.9 is that the outperformance generation value and simulation time increase significantly as the trading period increases. The reason can be found by comparing the last two rows of Tables 7.4 through 7.8. The normality of the trading assets improves as the trading period increases. The skewness approaches zero (normal distribution) and the kurtosis approaches 3 (normal distribution) as the trading period increases. We can also observe this effect by comparing Figures 7.11 and 7.14, Figures 7.12 and 7.15, and Figures 7.13 and 7.16.

The reason for the improvement of the normality of the trading assets lies in the accumulation effect. A trading period (hour, day, week, month, or year) consists of a section of trading intervals. Therefore, the effectiveness of the MV-A and the EUT-S portfolios are in Tables 7.4 through 7.8. The result of these improvements is that the EUT-S method requires many more generations to outperform the MV-A portfolio.

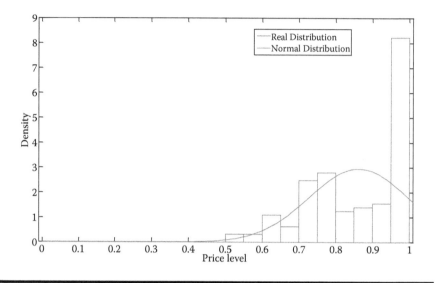

Figure 7.12 Frequency distribution of CfDC price (hourly).

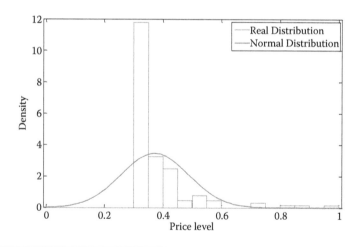

Figure 7.13 Frequency distribution of CfDS price (hourly).

Although each $\tilde{p}_{m,t}$ alone does not necessarily follow a normal distribution, the aggregation of $\tilde{p}_{m,t}$ values contributes to the normality of the portfolio revenue. Therefore, the effectiveness of the MV-A model improves. This can be seen also by comparing the difference between the MV-A and the EUT-S portfolios in Tables 7.4 through 7.8. The result of these improvements is that the EUT-S method requires many more generations to outperform the MV-A portfolio.

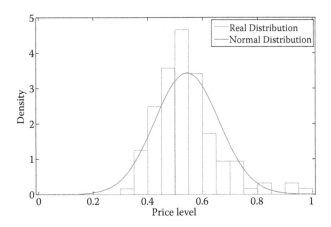

Figure 7.14   Frequency distribution of DA price (daily).

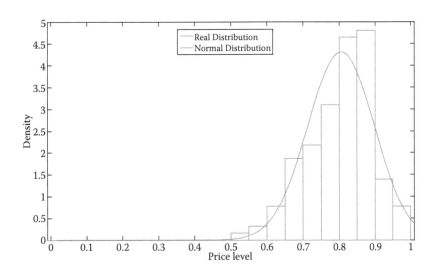

Figure 7.15   Frequency distribution of CfDC price (daily).

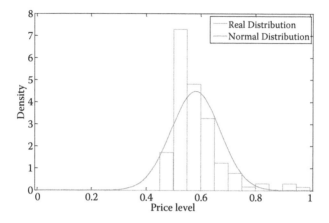

Figure 7.16   Frequency distribution of CfDS price (daily).

References

P. Artzner, F. Delbaen, J. M. Eber et al. 1999. Coherent measures of risk. *Mathematical Finance*, 9, 203–228.

D. W. Bunn. 2000. Forecasting loads and prices in competitive power markets. *IEEE Proceedings*, 2, 163–169.

M. Carrión, A. B. Philpott, A. J. Conejo et al. 2007. A stochastic programming approach to electric energy procurement for large consumers. *IEEE Transactions on Power Systems*, 22, 744–754.

A. J. Conejo, R. Garcia-Bertrand, M. Carrión et al. 2008. Optimal involvement in futures markets of a power producer. *IEEE Transactions on Power Systems*, 23, 703–711.

A. Eichhorn, N. Growe-Kuska, A. Liebscher et al. 2004. Mean-risk optimization of electricity portfolios. *Applied Mathematics and Mechanics*, 4, 3–6.

D. Feng, D. Gan, J. Zhong et al. 2007. Supplier asset allocation in a pool-based electricity market. *IEEE Transactions on Power Systems*, 22, 1129–1138.

D. Feng, Z. Yan, J. Østergaard et al. 2010. Simulation embedded artificial intelligence search method for supplier trading portfolio decisions. *IEEE Transactions on Generation, Transmission, and Distribution*, 4, 221–230.

Bibliography

Z. Bodie, A. Kane, and A. Marcus. 2005. *Investments*, 6th ed. New York: McGraw Hill.

Z. Hu, L. Yang, Z. Wang et al. 2008. A game theoretic model for electricity markets with tight capacity constraints. *International Journal of Electrical Power and Energy Systems*, 30, 207–215.

Index